W9-AFB-752

Killer Bees

Killer Bees

The Africanized Honey Bee in the Americas

Mark L. Winston

Harvard University Press
Cambridge, Massachusetts
London, England
1992

Printed in the United States of America
10 9 8 7 6 5 4 3 2 1

This book is printed on acid-free paper, and its binding materials have been chosen for strength and durability.

Library of Congress Cataloging-in-Publication Data

Winston, Mark L.
Killer bees : the Africanized honey bee in the Americas / Mark L. Winston.
p. cm.
Includes bibliographical references and index.
ISBN 0-674-50352-X (acid-free)
1. Africanized honey bee. 2. Africanized honey bee—America. 3. Africanized honey bee—Control. 4. Africanized honey bee—Control—America. I. Title.
SF538.5.A37W56 1992 91-33113
638'.12—dc20 CIP

For Sue and Devora,
still with love

Contents

Preface

My objective in this book is to chronicle the biology and impact of a fascinating insect, the Africanized—or "killer"—bee. The appearance of this bee in South America has resulted in an entomological phenomenon unparalleled in the history of exotic pests. Countless insects, both beneficial and detrimental, have been introduced to locations throughout the world; but the Africanized honey bee (as it is commonly referred to in the United States) stands out because of the rapidity of its spread and the economic devastation it has wrought. The enormous amount of media attention to this insect has strongly influenced the way entomologists practice their normally meticulous and detail-oriented craft. Uncharacteristic media battles take place between scientists who are well prepared for academic disagreements but not for airing disputes in public. Research into the biology and management of these bees is set against a background of controversial policy decisions, action plans, government programs, and highly charged competition for funding between government and university researchers.

Indeed, the politics of the Africanized honey bee, and the media attention to it, have caused us to lose sight of the unprecedented success story of an introduced species that is elegantly preadapted to its new environment. The bee is so well suited to tropical life that we have not been able to devise a way to stop or even slow its spread. We marvel at its success in the wild, even as we struggle to blunt its impact on beekeeping and the public.

The Africanized bee and I intersected in the mid-1970s, when it was becoming increasingly clear that this bee posed a serious threat to North America. A young biology student in search of a focus for my career, I had studied single-celled organisms in the laboratory of the eminent microbiologist Lynn Margulis at Boston University, and had spent three years as a marine biologist in Woods Hole, Massachusetts—first as an employee of the Oceanographic Institute and later as a graduate student in a Master's degree program at the Marine Biological Laboratory.

In the fall of 1975 I arrived at the University of Kansas to pursue a doctorate in entomology. I wanted to study insects in the tropics, and the Entomology Department at the University of Kansas had a reputation for allowing students to do exactly that. My supervisor, Orley "Chip" Taylor, mentioned that he had just received a federal grant to study the biology and potential management of the Africanized honey bee and was assembling a team to work in French Guiana, on the northeast corner of South America. I knew instantly, with no questions or hesitations, that this assignment was what I was looking for, although it took almost a year to get to French Guiana and start work.

The first team in French Guiana consisted of myself, Gard Otis, David Roubik, their wives and children, and of course our leader, Chip Taylor—who periodically came down with fresh T-shirts, mail, equipment, and lots of advice and enthusiasm. Calling ourselves the killer bee team, we were an unusual group of students: when we started, none of us had had any experience with honey bees. This deficiency turned out to be a real advantage in that we had none of the preconceptions that researchers with a more traditional background in apiculture might bring to the study. We treated Africanized honey bees as highly successful feral organisms in a complex ecosystem and set out to determine why they were doing so well.

French Guiana is a remote, largely unpopulated protectorate of France. The coastal region where we worked is a mix of savannah and rain forest; it has a dry season that lasts about eight months, during which it rarely rains, and a wet season of four months' duration, when it rains daily. The country is best known for the penal colony of Devil's Island, where Alfred Dreyfus and the infamous escapee Papillon were ensconced (the prison is now a museum). French Guiana is home to one of the few remaining Foreign Legion outposts in the world, and to a high-technology satellite-launching facility that has sprouted in Kourou, the town that used to support the prison.

The Africanized bees had recently arrived in French Guiana, and we wanted to study them during their initial colonization and subsequent spread. Also, the small human population made it easy to find remote apiary sites where we could make mistakes without anyone's getting stung if we angered the bees. In this tropical backwater, with its almost pristine habitat, amid the occasional roar of missiles as they launched and crashed and the increasing buzz of feral honey bees, we began to unravel the story behind the success of the Africanized honey bee in the Americas.

In this account I describe first the biology of the Africanized bee, from my own experience and from reports in the literature, emphasizing the aspects that have been most relevant for its proliferation. In the second part of the book I delve into economic impact and agricultural policy, describe various conflicts and disagreements that have developed concerning this unusual bee, and provide recommendations for dealing with the problems that will arise as the Africanized bee colonizes much of the southern United States. I have written subjectively, stressing what I think is most important; other writers might have a different perspective and come to different conclusions.

In the writing of this book I have incurred many debts. I am deeply grateful to Susan Katz and Keith Slessor, who provided endless encouragement and critically read the entire manuscript. I am also indebted to Ron Einblau, Glenn Hall, Harriet Lemer, and Eric Mussen, each of whom read sections of the manuscript and gave much helpful advice; to James Tew, who provided some of the economic data in Chapter 8; and to Debra Swain, who ably conducted background research on numerous topics. I appreciate the willingness of various colleagues to send me reprints and to share unpublished manuscripts and other materials. This book is richer and more complete because of their generosity.

Linda Barkley kindly gave me permission to quote extensively from one of her letters. The figures and chapter opening sketches were drawn by Elizabeth Carefoot; I continue to enjoy and appreciate her enthusiasm, and marvel at her technical excellence.

I am grateful for support and encouragement over the years from Simon Fraser University—especially for my appointment as a University Research Professor, which gave me the time and freedom to complete this book. My thanks go also to the following publishers and individuals for permission to redraw figures from their publications: *American Bee Journal,* the International Bee Research Association, Pergamon Press, and O. R. Taylor. And I am grateful to the editors and staff at Harvard University Press, particularly Linda Howe, Angela von der Lippe, and Vivian Wheeler, for assistance at all stages of the writing and publication of this book.

Finally, I express deep appreciation to my colleagues Gard Otis, Chip Taylor, and the other members of the French Guiana killer bee team—Helen Hasbrouck, Penny Kukuk,

David Roubik, and Ellen Winchell. My own research would not have been possible without their collaboration and support, and our shared experiences and enduring friendship have been the most gratifying part of my involvement with the Africanized honey bee.

I

Biology and Habits

1

The Creation of a Pop Insect

The Africanized bee is the pop insect of the twentieth century. Media star of tabloids, B movies, and television comedy, it has been nicknamed the killer bee. The media have largely ignored the intriguing natural history behind this insect's proliferation and have paid scant attention to its economic impact. Rather, attention has focused on shock stories and jokes, bad puns, and lurid tales of death by stinging. As a result, the public's impression of the Africanized bee goes far beyond its natural significance, and the normal fear in which people hold bees has become exaggerated to a ludicrous extent. Small wonder: books and movies depict killer bees attacking cities, forcing the evacuation of all humans from Latin America, and destroying nuclear missile sites.

The fearsome image of Africanized bees starts with their nickname. These insects were not always called killers; early press reports referred to them as "African" bees, and a bit later as "Brazilian" bees. It was not long before the media began experimenting with new names, however; by 1965 *Time* magazine (in its issue of 24 September) was calling them "Killer Bees," and later the "Bad Bees of Brazil" (12 April 1968). By 1972 even the venerable *New York Times* was using the title "Aggressive Honeybees" (22 January), and in a headline on 15 September 1974 announced that "The African Killer Bee Is Headed This Way." The term had caught on, and the killer bee was launched on its path to stardom.

Indeed, the headlines about these bees declared that readers should not expect much in the way of substantive and balanced reporting. The 1965 *Time* article was entitled "Danger from the African Queens," and the situation went downhill from there. A few other gems have included "Those Fiery Brazilian Bees" (*National Geographic,* April 1976), "Stalking the Killer Bee" (*Boston Phoenix,* 29 March 1977), "Savaged and Stung: The Killer Bees" (*Rolling Stone,* 28 July 1977), and "Invasion of the Killer Bees; Really, They're Coming" (*Newsweek,* 6 April 1987).

The overripe phraseology of these headlines has been matched by the catchy lead-ins and first paragraphs of the stories themselves. The print media seem to lack faith in the content of their reporting; they frequently rely on snappy phrases, militaristic terminology, and sensationalist writing to drag the reader into their Africanized bee stories. The following are notable examples of this genre.

> Like an insect version of Genghis Khan, the fierce Brazilian bees are coming. Millions of them are swarming northward . . . liquidat-

ing passive colonies of native bees in their path (*Time,* 18 September 1972)

Much like the monster creations of science fiction, the northward swarming of vicious African honeybees . . . (*New York Times,* 15 September 1974)

The bees came on us like a squall. At first we felt only the warning, the pelting of a few sentinels against our protective veils. Then, as we drew closer to their hives in the equatorial Brazilian bush, the torrent broke . . . The bees seemed possessed by rage. (*National Geographic,* April 1976)

The truth is they're not as monstrous as they're portrayed in horror movies. They won't carry off your kids or attack you unprovoked. And they're so TOUGH they may actually help the American honey industry. But don't underestimate their mean streak. They've already KILLED hundreds of people, STINGING some thousands of times. The slightest jostle is enough to send them into a VICIOUS FRENZY. And now, they are heading this way. Not for nothing are they called THE KILLER BEES. (*Philadelphia Inquirer,* 30 July 1989).

Bad as the printed media reports have been, books and movies have been even worse. There is no pretense of accuracy, and imagination has taken over to bring us wildly exaggerated horror stories that play on our innate fear of stinging insects. A prime example of an overplayed and overdone "Bee" grade movie is *The Swarm,* a fictionalized tale of a giant killer bee swarm that invades the United States. The movie opens with the bees destroying a nuclear missile site in Texas, then overturning trains and attacking children on their way to Houston. They leave that city in flaming ruins before being destroyed by the U.S. Army Corps of Engineers: in a brilliant strategic move, the queen and her workers are attracted to an oil slick in the Gulf of Mexico by foghorns that imitate the mating call of the drone bee. The army lobs mortar shells into the oil, exploding it and the bees in a fireball that saves North American civilization from the onslaught of these ferocious monster insects.

The beekeeping community was in turmoil as the 1977 release date for this movie approached, for we all expected a substantial negative backlash against bees and beekeeping. We need not have worried; *The Swarm* was a flop and closed within days of its highly publicized opening.

A personal favorite of mine is J. Laflin's 1976 book entitled *The Bees*. The back cover proclaims it a "terrifying novel of natural violence" and goes on to describe the beginnings of "the war between man and bee." By the end of the book, millions have died and almost all of Latin America has been evacuated to the north. The U.S. government establishes a think tank deep in a protected bunker to come up with a solution, which turns out to be Operation Cold Front—hundreds of miles of pipe carrying refrigeration liquid to kill the bees by chilling them. For some reason, only the bees from temperate climates are killed, and the tropical killers continue to advance. But natural weather succeeds where Operation Cold Front has failed; a freak cold snap kills all the bees, and civilization is saved.

The best publicity about Africanized bees may have come from the satiric media, especially shows like "Saturday Night Live" that ran killer bee gags for years. Who can forget John Belushi and his Mexican bandits, dressed up like "Keeeler Beeees," demanding "Your pollen or your wife!" Killer bees have become part of our collective consciousness; bee costumes have become a standard Halloween motif, and no costume ball is complete without a few couples in wacky bee outfits.

We knew that the killer bee had indeed become America's favorite pop insect when Killer Bee Honey appeared on the market a few years ago. This product was marketed by a journalist (call him Ed), who has made a minicareer out of the bees. When our team was working in French Guiana in 1976,

Ed first visited us as a reporter for *Rolling Stone;* since then he has published many articles about Africanized bees. Like many of the reporters who came to French Guiana, Ed was impressed with his own bravery at putting "his life on the line to learn the *truth* about the killer bees," as he modestly put it in his article's headline. One of our favorite journalist-baiting routines in those days was passing out samples of honey from our hives and suggesting that someone could make a "killing" by bottling and marketing killer bee honey. Ed took this idea seriously. About a year later, just before Christmas, Killer Bee Honey hit the market. It sold for almost a dollar an ounce and came with a brochure that enjoined the consumer: "As you taste this honey, remember the lives it has cost. And then enjoy it. If you can." Ed went around the country in a beesuit and veil, promoting the product, but the novelty quickly wore off. One food critic described the honey as having "the taste of molasses and silage or hay in a country barn." Killer Bee Honey was a failure.

All the publicity about Africanized bees, from Killer Bee Honey to tabloid journalism, may actually have helped the beekeeping industry. The movies, books, and press reports have brought bees to the public's attention, and beekeepers have capitalized on this notoriety to make their point that bees are a critical part of North American agriculture, particularly with regard to crop pollination. We can expect a new wave of publicity as the bees spread through the southern United States, but let us hope that the scare stories have run their course. It is time for responsible journalism to take over; we need sound, accurate information about bees in general, and especially about Africanized bees. Beyond that, the factual story of the Africanized bees is as riveting as the lurid sting tales. Frightening the public is easy, informing it is more difficult; but it is information, not horror, that is needed now. The media created and exaggerated the killer bee monster; it is time to reduce it to its true insect size.

2

Arrival of the Bees

In 1976, as the killer bee team set up its operation, the Africanized honey bee had been in South America for twenty years. The first swarms were just arriving in French Guiana from the south. What we subsequently saw essentially repeated events that had occurred in Brazil in 1956 and continue even today, as the bees migrate and spread into new habitats.

Our presence in South America, and indeed all of the events associated with Africanized bees, began with the supposedly harmless introduction of a few honey bee queens from Africa into southern Brazil. Honey bees are not native to either North or South America, and the bees being used in

Brazil in the 1950s originated in Europe. These European bees, although reasonably gentle, were not good honey producers in tropical and subtropical climates and were particularly poor for beekeeping in the Amazon Basin. Warwick Kerr, a Brazilian geneticist, was asked by his government to initiate a program to import and breed bees more suited to the Brazilian habitat; for stock he naturally looked toward Africa, the original habitat of tropical honey bees.

Kerr and his team were intrigued by reports from South Africa claiming tremendous honey crops, including one article in a 1946 issue of the *South African Bee Journal* which reported a record 257 kilograms of honey produced in a single year from one colony, and annual averages of 70 kilograms per colony. The Kerr group knew that African bees had a reputation for being highly aggressive, but reasoned that they could cross African with European bees to produce a hybrid with the gentle European characteristics but the supposedly high honey production of the African bees. Consequently, Kerr went to Africa to collect live, mated queens that he planned to bring to Brazil to initiate this breeding program.

Kerr traveled throughout eastern and southern Africa, examining colonies and selecting queens for shipment. Most of his first shipments died before reaching Brazil; only one queen of 41 collected from East Africa survived. This queen was from Tanzania, where Kerr had found the bees to be unusually defensive. He continued on to South Africa, receiving 12 queens from E. A. Schnetler, the beekeeper whose colonies had set the honey-production record, and another 120 from W. E. Crisp, a commercial beekeeper who was president of the South African Beekeepers Association. Of these queens only 54 survived shipment and were introduced into Brazilian colonies, and 35 of these were selected for further testing and breeding.

Kerr believed these colonies, now headed by African

queens, to be among the most productive he had ever seen. In 1957 they were moved to a eucalyptus forest in São Paulo for further evaluation. Their entrances were fitted with queen excluders, which are sheets of screen through which the workers can pass but the larger queens cannot. These excluders were supposed to keep the colonies from escaping into the wild by confining the queens, but at some point a local beekeeper apparently removed the screens, and 26 of the colonies swarmed into the forest. *Swarming,* or colony reproduction, occurs when the majority of workers leave the nest with a queen and search for a new nest site, leaving behind some workers and another queen to continue the original nest. The São Paulo swarms included the queen from Tanzania and 25 queens from the Transvaal province of South Africa. There is evidence that additional African queens were reared from the remaining colonies and distributed to Brazilian beekeepers—which may have been a more important source of African bees than the well-publicized escape.

The escaped swarms, along with any progeny of distributed queens, formed the nucleus of a feral population which has since spread at tremendous speed and density through most of South America and Central America (Figure 1).The expanding front of the bees has been moving 300 to 500 kilometers a year, with even moderate-sized local populations reaching densities of 6 colonies per square kilometer (densities as high as 108 colonies per square kilometer have been reported). David Roubik, one of the original killer bee team members who is currently at the Smithsonian Tropical Research Institute in Panama, estimates that there are currently one trillion individual Africanized bees in Latin America, which would make up 50 million to 100 million nests—and these estimates probably are conservative.

Our team witnessed the initial impact of these bees as they arrived in French Guiana and neighboring Surinam. The sce-

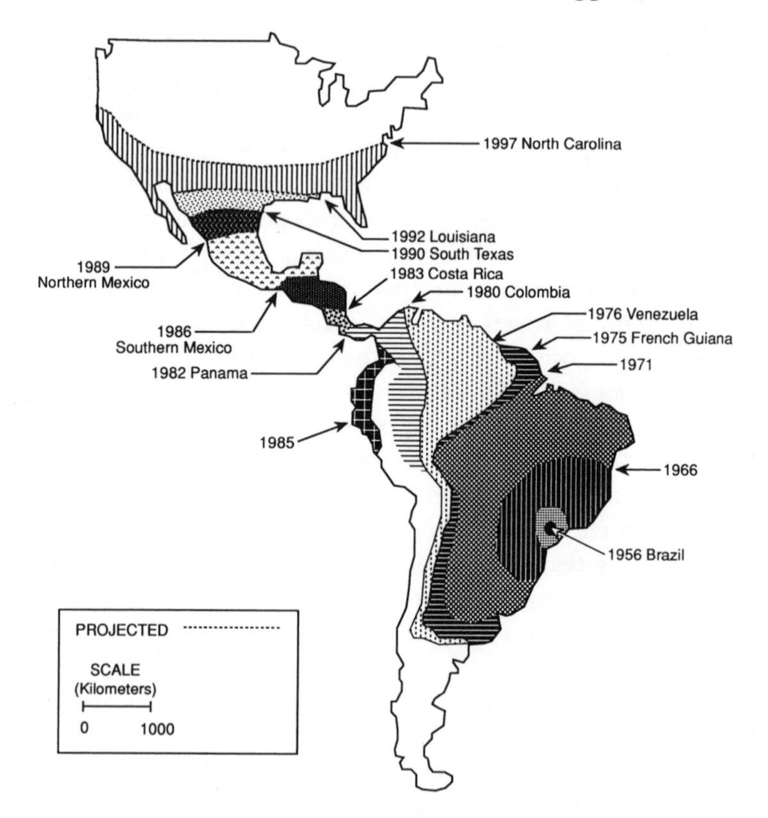

Figure 1. Actual and projected rate of spread of the Africanized honey bee in the Americas.

nario was identical to what had previously occurred in Brazil, and would occur repeatedly throughout Latin America as the bees expanded their range. Densities of feral colonies remained low during the first year or two following colonization; in fact, we even had difficulty finding enough bees to study. Then, population growth exploded. Suddenly everyone knew of feral colonies, and beekeepers were overwhelmed with abrupt changes in their colonies' behavior. The number of stinging incidents increased sharply, so that beekeepers could no longer maintain their colonies near people or livestock. Honey production diminished to near-zero levels due to excessive swarming, *absconding* (when the entire colony forms a swarm and abandons the nest, leaving only the comb behind), and the reluctance of most beekeepers to attempt even simple management in the face of massive stinging. Clearly, managed colonies were becoming "Africanized," taking on the traits of the feral bees through a combination of queens mating with feral Africanized drones (male bees) and Africanized swarms entering managed colonies and taking them over. Most beekeepers did not understand this process of Africanization, much less have the training or resources to respond, and it was not surprising that the vast majority of hobby and even commercial beekeepers soon abandoned their craft and their colonies, often leaving hives in their apiaries and not going back.

The few commercial beekeepers who remained in the business had to institute radical changes in their management procedures; beyond that, the fun had gone out of beekeeping. Apiaries had to be moved far from people and livestock, to remote areas where honey production often was diminished from that of the agricultural regions with blooming crops that beekeepers prefer. Apiarists had to learn to cope with excessive stinging, swarming, and absconding, and had to accept less honey production from each colony.

My 1977 visit to an apiary in Surinam that had recently become Africanized was typical. A local schoolteacher maintained about twenty colonies in an isolated grove out in the country, a few hundred meters away from a number of small farms. He abandoned his colonies shortly after my visit, but at that time he was still keen on beekeeping and eager to convince me of the advantages of these bees. Even so, we parked about a half-kilometer from his beeyard, put on two layers of clothes under our bulky beesuit coveralls, and carefully secured our veils and gloves to leave the bees no room to enter. Then we lit the largest smokers I had ever seen, bellows-like instruments that burn burlap, old sheets, cardboard, dried cow patties, or whatever is available to generate smoke to pacify the bees. Only then did we approach his colonies, and I should have been warned by these elaborate precautions.

Merely walking toward the colonies elicited a massive response on the part of the bees, so that the situation was out of control before we smoked and opened our first colony. Bees were everywhere, banging into our veils and helmets with such ferocity that we could barely hear each other and stinging through our layered clothing. It was a hot, humid day, and the combination of sweat, noise, and stings forced us to retreat after examining only a few colonies. The bees followed us all the way back to the car, and we had to keep our equipment on until we were far out of their stinging range. As we drove off, we could see the farmers swatting at bees and two of their cows were being stung; we had to stop and move the animals farther away to safety.

Fortunately, no one was seriously hurt, but this incident was not unusual. There have been frequent deaths of livestock that are fenced or chained too close to Africanized hives, and there have been some human fatalities due to massive numbers of stings. On this occasion I received over fifty stings in just a few minutes, in spite of my heavy beekeeping armor, and I could only imagine what a full day of working with these bees would be like.

These events have been repeated throughout South and Central America, and undoubtedly will occur to some extent in North America as the bees move through Mexico into the United States. The situation is not all grim, however, because the arrival of Africanized bees, which has disrupted beekeeping all over Latin America, has been followed by familiarization with these bees, and selection and breeding of more tractable Africanized bees. While the arrival of the bees has consistently disrupted beekeeping country by country, beekeeping generally rebounds within five to ten years (although not to the levels found prior to Africanization). Also, stinging incidents diminish as the public and government agencies learn how better to deal with the presence of a new type of stinging insect. It is important to remember that people have coexisted with these bees in Africa since the first humans evolved on that continent, and beekeeping with appropriate management procedures has been viable there. Nevertheless, the Africanized bee remains a serious problem in the New World, because of its continuing adverse effect on beekeeping and the occasional dramatic stinging incidents it causes.

In hindsight, the importation of Africanized bees should not have taken place—or at the least, stock should have been properly selected, bred, and tested prior to importation. The bee's reputation as a good honey producer has proven unfounded, due to a combination of high swarming and absconding rates and the unwillingness of most beekeepers to perform even minimal management in the face of serious stinging problems. By any economic, agricultural, public, or political measure this importation was not desirable. In a biological sense, though, the bee has been highly successful, spreading at high rates and forming a dense feral population which may be having considerable impact on resident bees. Ironically, the characteristics that have proven deleterious for beekeeping ideally preadapted the African honey bee for a feral existence in South America.

3

Temperate and Tropical Honey Bees

The tropical origin of the Africanized honey bee is the key to understanding its success in the New World. Honey bees exhibit a wealth of variation throughout their original geographic range, with local races showing superbly fine-tuned adaptations to their particular environments. A *race* is a group of individuals with similar characteristics, usually from a distinct geographic region. Individuals of different races are members of the same species, and thus are related closely enough so that they can interbreed. The diverse types of honey bees can be broadly divided into two groups, the temperate-evolved European and the tropical-evolved African races.

Honey bees are not native to the Americas, but the European bees transported by man to the New World have proven to be almost perfectly preadapted to more temperate regions, performing admirably in beekeeping contexts but also establishing successful feral populations throughout southern Canada, the United States, Mexico, and southern South America. These temperate-evolved bees have not done well throughout the more tropical habitats of Latin America, however; European bees have never established any substantial feral presence in the tropical Americas. In contrast, we have seen that the tropical-evolved African bees moved quickly through tropical South and Central America, establishing enormous feral populations; their spread southward stopped when they reached the more temperate regions of Argentina.

The different patterns of climate, resource distribution, and predator abundance in Europe and Africa have been particularly important in molding the traits of temperate and tropical honey bees. In temperate climates, cold winter conditions produce dramatic changes in honey bee biology. The bees cluster tightly inside the nest, using stored honey as an energy source to generate heat by shivering. Brood rearing is largely curtailed, there is no foraging, and workers show extended life spans due to their quiescence. Winter has a very different meaning for tropical honey bees. Temperature differences between seasons are minimal, and it is rainfall that determines seasonality because of its effects on flowering and on nectar and pollen production. In Africa the dry season is the dearth period, while in South America the wet season is the time of reduced flowering; but in both situations the bees show differences in their biology which are as dramatic as the differences between winter and summer in temperate climates. The most notable difference is the tendency of tropical bees to abscond during the dearth season—in other words, to abandon their nests and move long distances in search of better resources.

While climate and patterns of resource distribution have

had profound effects on the evolution of temperate and tropical bee races, predation has also been a strong selective force. The large number of predators in tropical habitats has had considerable impact on tropical bees, particularly in the evolution of defensive behavior. The feistiness of Africanized bees, which has earned them their "killer" moniker, has its roots in millions of years of heightened predator pressure. Attackers such as ants, honey badgers, and humans have been a strong evolutionary influence on the predisposition of tropical bees to sting.

Honey bees originated in tropical Asia, where even today the largest number of species is found. These early honey bees and most of their Asian ancestors built open nests external to cavities, consisting of vertical wax combs with hexagonal cells covered with layers of worker bees for protection. At some point, however, a line of honey bees diverged from this ancestral type and began nesting inside cavities. It was from this line that the western honey bee species *Apis mellifera* evolved, with its European and African races.

These cavity-nesting honey bees migrated to Africa a few million years ago, colonizing the colder climates of Europe somewhat later. The separation of the Asian and Afro-European groups of honey bees into separate species may have been a more recent event, occurring between two million and three million years ago and resulting from separation of the two regions by glaciation. The continued divergence of *Apis mellifera* into distinct races was undoubtedly influenced by the advance and retreat of glaciers, which resulted in temporarily isolated populations under strong selective pressure to evolve behaviors to survive climatic extremes. Until modern times honey bees were not found anywhere in the western hemisphere; movement of bees by European settlers for beekeeping purposes has resulted in worldwide distribution. The major races, or subspecies, of European bees that have been imported to the Americas include the Italian bee (*Apis mellifera ligustica*), the carniolan bee from Austria and Yugoslavia (*Apis mellifera carnica*), and the German black bee (*Apis mellifera mellifera*). The African subspecies that was brought to South America was *Apis mellifera scutellata*, sometimes called the East African bee.

The original habitat of the western honey bee extends from the southern tip of Africa through savannah, rain forest, desert, and the mild climate of the Mediterranean before reaching the limit of its range in northern Europe and southern Scandinavia. It is not surprising that numerous races have evolved with widely divergent characteristics adapted to particular habitats. Beekeepers have taken advantage of this variety by importing bees of different races all over the world, attempting to match their local conditions with the areas of origin for each race, and cross-breeding races to produce unique combinations of characteristics that might be better suited to new habitats. The genesis of the Africanized bee problem was just such an importation.

The extremely social nature of a honey bee colony involves intricate interactions among a large number of individuals, so that the sum of colony functioning is much greater than any one individual's performance. Coordination of the activities of up to fifty thousand insects is a complex task, but the social interactions of so many individuals has provided considerable flexibility for honey bee societies to survive and evolve in both temperate and tropical environments.

The differences between the African and European honey bee races which have evolved in the two habitats involve sets of parallel traits. Temperate bees build colonies in large nests which store a considerable quantity of honey, rarely abandon their nests, and reproduce relatively rarely. Tropical colonies put their energy into reproduction rather than honey production; they construct small nests, store relatively little honey, reproduce frequently, and readily abscond. Also, everything from worker behavior to colony growth and reproduction happens at a much faster pace in tropical bees; individuals seem to work harder and die younger.

Indeed, my first visit to an Africanized bee colony impressed me with the fast-paced lifestyle of these bees. I arrived in French Guiana late one night in the summer of 1976, and the team went out early the next morning to see our first Africanized bees. When we opened up the first colony, I was still keyed up from travel and jet lag; my adrenalin was flowing, partly from the excitement of beginning my research but also because I was fearful of a vicious attack from the supposed killers. To my surprise, the bees did not attack us, but seemed to be even more nervous and highstrung than I was. Although this was a small colony, the activity level inside was astounding, with worker bees running in waves across the comb and down through the hive. Even the queen, who in European colonies is slow and ponderous, was, in beekeeper terminology, "runny."

As we continued to examine colonies, we began to realize that these observations might be the focus of our first research. Beekeepers had frequently commented on how quickly a small colony of Africanized bees can grow and swarm, often before proper management procedures can be initiated. We decided to investigate the interaction between growth rate, activity level, and life span of individual bees, along with the rapid colony growth and reproduction which obviously is a major component of Africanized bee biology.

The most apparent developmental difference between Africanized and European bees is the rapidity of development of the Africanized brood. Honey bees go through four developmental stages: egg, larva, pupa, and adult. The eggs are laid in the bottoms of cells, one egg per cell, by the queen. The eggs hatch into young larvae after about three days. The larval stage is the feeding time, when the bee gains an enormous amount of weight and grows tremendously in size. For example, larvae of worker bees (all of the workers are female) gain nine hundred times the weight of the egg in only four to five days. This growth is fueled by the nectar and pollen that is collected from flowers and processed by the adult workers; these are the only food sources required by honey bees. When larval growth is complete, the adult workers cap the cells with wax; the larvae inside metamorphosize (change form) into pupae, and then finally into adults, just as a caterpillar changes into a pupa and then into an adult moth or butterfly. When this transformation is complete, the young adult chews out of its cell and finishes developing during the next few days.

The development time of each stage is slightly shorter for Africanized bees. Total time from egg laying to emergence of adult workers is about 21 days for European bees and 18.5 days for Africanized bees. The Africanized worker bees also are about 10 percent smaller and 33 percent lighter than European bees (62 milligrams as opposed to 93 milligrams, on average).

A consequence of these developmental differences is that Africanized colonies produce adult workers at a more rapid pace than European bees, although each worker is smaller. It occurred to us that another result might be differences in adult life span and in the ways that workers of various ages allocate their work performance. Honey bee workers can make two types of adjustments that are important for colony functioning.

1. Younger individuals tend to perform tasks within colonies, such as brood rearing, cleaning, and nest construction. As they age, workers shift to outside tasks such as guarding and particularly foraging, until they die. Workers can alter this temporal work ontogeny considerably, however, to meet colony requirements. For example, older bees can rear brood if colonies require it, and very young bees can forage. This

ability to adjust labor schedules is an important component of a colony's ability to respond to changing environmental conditions, such as the discovery of an abundant resource, predation, or an unpredictable change in the weather.

2. Workers become relatively quiescent during the cold winters, thereby extending their life span from the 25 to 40 days typical of summer bees to 140 days or longer during the winter.

Although these characteristics were well known for European bees in temperate climates, virtually nothing was known about how Africanized bees might utilize such traits to maximize colony survival in tropical habitats. We began our study of adult life history characteristics by examining how long worker bees live during different seasons, which is a conceptually simple but pragmatically tedious task. Newly emerged bees are not fully developed, and it takes about twenty-four hours following emergence for their external skeleton to harden. Because the sting apparatus also is soft and weak, these baby bees cannot sting, and so can be easily handled and marked. We were able to mark each emerging bee with a small round tag glued onto the bee's back and coded with colors and numbers that could be read from the label to identify each bee. We marked many thousands of bees in French Guiana, and we would then put them into a coffee can, drive out to the apiary sites, and shake batches of a hundred at a time into colonies, which readily accepted these youngsters. Subsequently, we examined each colony weekly to note the numbers of surviving workers, going through colonies at least twice during each inspection to ensure that no living workers were missed.

The results of these studies yielded valuable information concerning worker life history and the dynamics of colony growth, and stimulated other lines of research which are still yielding profitable results. The most striking finding was how short-lived the Africanized workers were. Average survival times for European workers during the summer generally are between 20 and 35 days; Africanized workers show average life spans of only 12 to 18 days during the equivalent dry season in French Guiana, and slightly higher spans during the wet season, about 20 days. Winter results are equally dramatic; one study from Poland showed that Africanized workers live only 90 days during the winter, as opposed to the 140-day average for European bees.

The shortest average life span ever recorded for honey bees came from studies of the first Africanized workers to emerge in colonies after swarms were established in new nests. Their abbreviated 12-day life was undoubtedly due to the tremendous amount of work needed to construct a new nest, rear brood, and forage while the colony population was still low. Curiously, workers of European races emerging in new colonies do not show diminished longevities, which is consistent with the slow but steady growth that is more characteristic of European colonies.

The combination of these survivorship studies suggested that the integration of work tasks and life span may be substantially different for European and Africanized bees. That is, it seemed that European bees are relatively lazy, work at a slower pace, but live longer. Africanized bees, on the other hand, go full out during their short lives, accomplish more work in less time, but die younger.

These ideas took a number of years to ferment, and by the time they were ready for testing the killer bee team's base of operations had moved to Maturin, Venezuela. This cattle country had none of the international ambience of French Guiana. The Venezuelan government had built a research station for us, equipped with all of the comforts they imagined North Americans would want: air conditioner, television, and waffle iron! Unfortunately, the electricity and plumbing hardly ever worked, and we rarely were able to use these conveniences. The station was otherwise picturesque, located

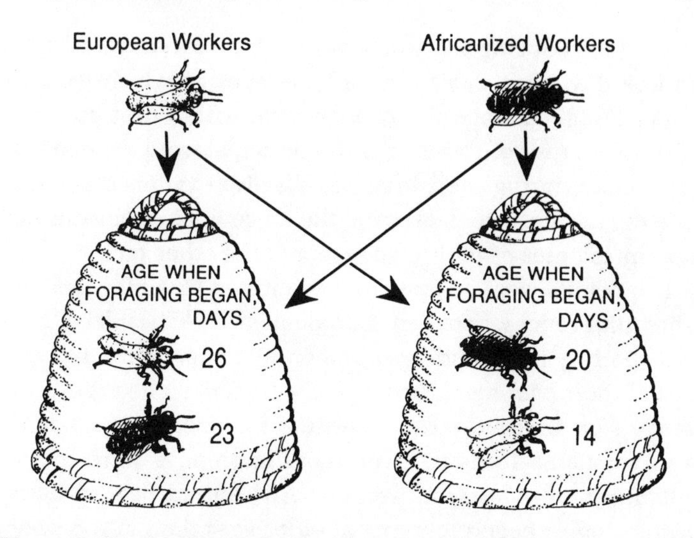

Figure 2. The experimental design used to cross-foster individual Africanized and European worker bees in colonies of both races.

in an old mango plantation, with hundreds of bee colonies under the large, stately trees.

My wife (Susan Katz), who was trained as a mammalogist, and I went to this unlikely setting during the wet season of 1980, to determine whether European bees were indeed lazier than tropical bees. We decided to try the experimental technique of cross-fostering, popular among psychologists but at that time rarely used by bee biologists. This approach is useful in separating environmental from genetic components of behavior, which can be accomplished by taking individuals suspected to have genetic differences and putting them in the same environment. With honey bees, we put newly emerged and marked Africanized and European worker bees into their own colonies and into colonies of the opposite race. We then looked for similarities or differences in life span and the age-based task of foraging in the two colony environments (Figure 2).

After introducing the young workers, we sat at the nest entrances day after day to record the ages at which workers began foraging, since the change from within-nest duties to the outside task of foraging is the principal work change that bees make during their lifetime. We also examined colonies on a regular basis to determine the longevities of our marked bees in colonies of their own race and the other race.

The results were striking and confirmed that life span and labor differences between European and Africanized bees have both genetic and environmental components. In colonies of their own race, the tropical Africanized bees began to forage significantly earlier than the European bees—at 20 days versus at 26 days. Even more impressive were the results from cross-fostered workers: Africanized bees in European colonies began foraging at older ages than in their own colonies, 23 days on average. More dramatically, the European workers began foraging at only 14 days of age in the Africanized colonies. The life span results paralleled the foraging age results.

Our explanation for these results is that the environment in colonies of the two races differs, and the differences in response of each race's workers in the same-race colony indicates a genetic component to life span and behavior. The colony environments seem to vary in the level of stimuli to foraging behavior, which are higher in Africanized colonies. Africanized workers therefore begin to forage at younger ages and also die earlier. The stimuli might include age distribution of workers, colony size, worker activity level, amount of brood rearing, and other factors. Whatever the stimuli, they are lower in European colonies, and these bees are older when they begin to forage. When a European bee is put into an Africanized colony, however, she is bombarded with a higher level of stimuli than would be found in colonies of her own race, so she forages and dies at an even younger age than the Africanized bees. The differences between European and Africanized workers in the Africanized colonies provide particularly compelling evidence of genetic variation between temperate and tropical bees, since the two kinds of bees showed different foraging ages when placed in the same colony environment.

Not only did this remarkable finding explain in part the differences in colony dynamics of European and Africanized bees, it also stimulated lines of research at laboratories around the world which have resulted in a new understanding of the genetic makeup and functioning of colonies. Honey bee colonies used to be considered fairly homogeneous units, with all the workers responding in similar fashion to colony needs. Queens, however, usually mate with about ten males (drones), thereby creating numerous subfamilies with the same mother but different drone fathers. From cross-fostering techniques similar to those we utilized with the Africanized and European bees, other researchers found that each subfamily shows different responses to colony conditions. For example, if a colony is short of pollen, only one or a few of the subfamilies will increase their pollen-collecting behaviors. These pollen-collecting lines, on the other hand, may not increase their nectar collection if colonies are short of honey, leaving that task to subfamilies which emphasize nectar collection. In this way colonies may have specialist squads to deal with particular problems, thereby increasing the overall ability of colonies to respond to the range of environmental perturbations found in nature.

As we began working with feral colonies of Africanized bees in French Guiana, it quickly became apparent that they were different from European bees in more than just the characteristics of the bees themselves; the nests of the temperate and tropical races also are different.

Honey bee combs are among the marvels of animal architecture, consisting of precisely constructed, back-to-back arrays of hexagonal cells, arranged in parallel series (Figure 3).

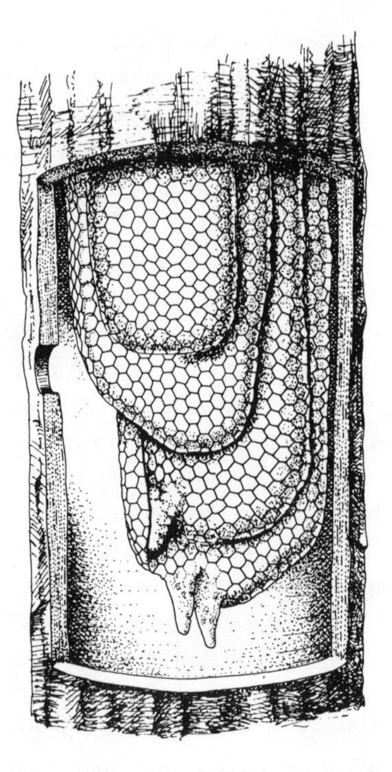

Figure 3. A typical honey bee nest inside a tree cavity. Four combs are shown; the average is five to ten.

Each comb, constructed from wax secreted by the workers, serves diverse colony functions, including those of pantry, nursery, and message center. The colony stores its honey and pollen and rears the brood in its cells, and the workers perform various communicative dances on the comb and transfer message-bearing chemicals which coordinate colony activities. The nest is exquisitely designed for these functions, but as we dissected more and more nests, we realized that variations in nest architecture between bee races reflect profoundly different adaptations for temperate and tropical living.

The importance to honey bees of a nest is apparent from the elaborate systems bees have evolved to choose nest sites, and from the time and energy they devote to nest construction. Honey bees select a nest site as the last stage of swarming, when the swarm has left the colony with its queen and clustered together, usually under an overhanging limb or in a snarl of branches. The swarm then faces a critical problem; it must quickly find a new nest site before the workers use up the honey they carry in their stomach when leaving the nest, or the swarm population will begin to dwindle as workers die. The swarm must also choose a site in which the new colony can survive and grow.

A swarm is faced with numerous potential sites to scout and choose from, and workers must reach a consensus on the preferred site before the swarm lifts off and moves to it. Once a swarm has settled into its interim clustering site, scouts are sent out almost immediately and begin to search for appropriate nests. When a scout has found a potential cavity, she spends considerable time examining it, evaluating cavity size, entrances, exposure to sun, draftiness, dampness, and other characteristics. Then she returns to the swarm and performs dances (see Chapter 5) that communicate the nest location and quality. Workers on the face of the swarm can "read" these dances, and soon more scouts fly out to examine the nest site. The swarm finally reaches a consensus when most of the scouts are dancing to the same location. At that point the scout bees perform a buzzing run, which causes the cluster to take to the air. The workers move with the queen to the new nest, following "guide" bees (which may be the scouts familiar with the new location).

Once the swarm has arrived at its final nest site, comb construction begins immediately. Many of the workers in the swarm already have begun producing beeswax for this purpose, by converting honey to wax in special abdominal glands. The wax is secreted as thin scales, which are manipulated by the worker bees into cells. An enormous amount of effort goes into this nest construction; the process of removing and manipulating each scale takes about 4 minutes, or 66,000 bee hours to produce the 77,000 cells that can be constructed from one kilogram of wax. Workers begin construction on the floor or side of a cavity, with perhaps two or three construction sites initially for each of the five to ten combs found in a typical colony. The completed nests of temperate-evolved European bees may last for many years, since the cells can be used over and over again to rear brood and to store honey and pollen. Also, these nests tend to be inside cavities for protection during cold winters.

Very little was known about tropical honey bee nests when we first arrived in South America. We examined our first Africanized nests in French Guiana, but were too busy with other studies to garner more than hints of any differences in nesting biology. It was not until the late 1970s, when we went to Peru and Venezuela to conduct nest examinations, that we made real progress.

One three-week period spent in the lowlands of Peru, while particularly productive, was the most arduous of all our research expeditions. We stayed with a missionary family who lived far down a freshly opened dirt road—a long, dusty, bumpy drive from the city of Pucallpa, on the Amazonian side of the Andes. These devoted missionaries were attempting to teach beekeeping to the new colonists in the area, and agreed to assist us in finding and dissecting nests if we would then transfer the bees to hives for the farmers to use for honey production. It was relatively easy to find nests here: the virgin forest was rapidly being cut and burned for agriculture, bringing down many nests in the process.

Nest dissection is physically demanding and time-consuming at best, and the living and working conditions at this site were like nothing we had previously encountered. We were in Peru toward the end of the dry season. It was unbearably hot, and the drinkable water supply was down to almost nothing; residents depended for drinking water solely on rainwater collected during the wet season, and the previous wet season had been unusually dry.

We were determined to find and cut up bee nests, however, and each morning we drove down the road and asked the local farmers if they had encountered any nests of the "abejas asesinas," the assassin or killer bees. Almost all of them knew of nest sites. To reach the nests, we had to carry axes, a chain saw, and measuring equipment along newly cut paths through the jungle, often walking for hours until we reached a nest. We would put on our beesuits, veils, and gloves, carefully take down and cut open the nest with the ax and chain saw, and remove the combs one by one for measurement. We would then strap the combs to frames and put them into a hive. The disoriented bees would eventually make their way back to their new home. They, of course, were not happy with having their nests torn apart, and we had to contend with frenzied bees while trying to make exacting scientific measurements. But by studying more than thirty such nests in one ten-day period during our stay in Peru, we learned a tremendous amount about how Africanized bees nest in the wild.

The most dramatic characteristic is the high proportion of nests found external to cavities, similar to the nests of the ancestral Asian species. These open nests may be located under branches, overhanging rocks, or buildings—and in some areas most of the wild nests are found in these situations. How such sites are chosen by swarms is not known; they may be interim swarm clustering sites at which workers begin constructing comb, and the swarm simply remains.

The diversity of exposed nesting sites which these bees will accept is astounding. We took nests from sewer manholes, old tires, rusting cars, empty oil drums, air-conditioning ducts, and assorted snarls of branches, limbs, and stinging plants in addition to the log cavities that are the better-known nesting sites for honey bees. These bees are everywhere; in French Guiana, we even had swarms come in our front window and settle in stored hives inside our house, undoubtedly attracted by the odor of previous bee tenants. We did not mind the intruders, however. Swarms are almost always docile, and we rarely experienced any serious stinging incidents while handling them.

Another impressive attribute of Africanized bee nests is their typically small size, generally one-third to one-half the size of an average European colony. We took out nests in Peru which were astonishingly small, sometimes taking up no more volume than a soccer ball yet containing old, dark comb which indicated that the colonies had been resident for a considerable time. Curiously, many of these nests were inside cavities with much greater volumes than the bees were occupying; the nests apparently were not limited by cavity size but by the preference of the bees for small nests.

Why might small nests and nests external to cavities be common for tropical bees and not for temperate bees? First, temperate-evolved bees require relatively large cavities with sufficient room to store the honey needed to survive winters, whereas tropical bees do not need to store massive quantities of honey. Second, bees in temperate areas maintain the colony temperature during the winter by clustering together, consuming stored honey for energy to generate heat. Winter survival in these regions requires not only substantial honey reserves but also a large worker population, and therefore a relatively large nest, inside a well-insulated cavity. Tropical bees, which do not experience a prolonged cold season, can manage with smaller nests and less honey, and the colony can survive outside a cavity. Also, external nests are more easily cooled, particularly if they are located in shady undergrowth. Finally, predation on honey bees is much more intense in tropical habitats, and small nests are more easily defended.

If a predator succeeds in destroying a nest, much less is lost by a small colony, which can leave the site and begin a new nest elsewhere.

Thus, nest sites seem to reflect adaptive differences between temperate-evolved and tropical-evolved bee races. Nest characteristics are not determined solely by the physical environment, however, but by colony functioning as well—particularly colony growth, reproduction, and absconding.

4

Seasonal Patterns, Swarming, and Absconding

Our killer bee team in French Guiana specialized in swarms. We were enthusiastic about almost anything we did back then, but swarms were special. Calls alerting us to swarms received our highest priority; anything else we were doing was put aside until the swarm was safely caught and settled in a new hive. We lived, breathed, and often literally were covered with swarms for over a year. In the process we learned a great deal about swarming, especially why it was such an important component of the Africanized bee's success. Beekeepers in general expend great effort to prevent swarming, because the worker populations of colonies that swarm are so severely reduced that they produce little or no honey. Our perspective was different: we spent countless hours waiting for swarms to issue and examining colonies before and after they did swarm, in order to determine why and when swarming would take place. We were particularly interested in ascertaining how Africanized bees could swarm so often and produce so many swarms during a single cycle.

Swarming is reproduction by colony division, whereby the old queen and a majority of the workers depart the colony to establish a new nest elsewhere. A new queen is reared to continue the original colony. Although swarming occurs mainly during the spring in temperate climates, preparations begin in the dead of winter, when colonies begin rearing their first worker brood. This early brood rearing is fueled by the vast quantities of honey and pollen stored in the nest during the previous summer, and it compensates for the gradual decline in adult population during the winter. By April the worker population has begun a dramatic increase, which climaxes in swarming. The time of swarming varies between regions, but usually peaks in May or June; years or locations with late springs show later peaks.

Direct preparations for swarming begin two to four weeks ahead of time, during a period when colonies are becoming congested because of the rapid growth of the worker population. For a reproductive swarm to issue, colonies must have produced or be capable of producing one or more new queens, so swarming preparations primarily involve the initiation of queen rearing. Once developing queens are present in colonies, the swarm can issue.

Queen rearing begins when the queen lays eggs into special downward-facing cells. Or workers may move queen-laid eggs from regular cells into the queen cells. Once the eggs in these queen cells hatch, workers feed the larvae specialized food called royal jelly, which induces the larvae to deviate from the usual worker development toward queen development. The larvae are sealed in their individual queen cells when growth is completed, and they then pupate and metamorphose into adult queens. Development from egg to adult queen takes about 16 days, and colonies generally initiate the rearing of 15 to 25 potential queens prior to swarming.

The first or prime swarm usually leaves when or shortly after the first queen cell is sealed, since the presence of a sealed cell ensures that at least one virgin queen will emerge to continue the original colony. The issuing of a swarm is a spectacular event. Workers first engorge with honey from their nest to provide food reserves until the new nest is established. Then, usually in the late morning or early afternoon, the workers run frenziedly back and forth in the nest, chasing, biting, and pulling their queen along with them. Suddenly a torrent of workers pours out of the colony entrance and takes to the air, driving the queen out as well. The deafening buzz of flying bees gradually subsides as the workers gather in a cluster with their queen, orienting to one another via the different blends of highly attractive chemicals which both workers and queens secrete. Scout bees then fly from the cluster to search for a new nest site. When a consensus is reached on the location of their new home, the swarm takes to the air and moves cross-country to settle into the new nest.

Left behind in the original colony are developing queens, brood, and stored honey and pollen. The first virgin queen emerges about a week later, frequently followed a few days after that by a smaller swarm called an *afterswarm*. The workers continue to permit one queen at a time to be released from her cell and issue with an afterswarm, until the colony's worker population is so depleted that no additional afterswarms are possible. At that point the remaining queens battle for supremacy until only one queen remains alive. That resident queen will mate, and the colony cycle begins anew.

Swarming occurs during the dry season in French Guiana, from July to February, which is the time of copious flowering in that part of South America. Gard Otis and I began our swarming observations early in July, when the colonies were still remarkably weak, and we feared we would have to wait many months before seeing any real swarm action. But by August colonies had filled their hives and the swarming season was in full swing. Fortuitously, we had put marked bees in the colonies for our worker life-span studies, and we were taking careful measurements of colony growth. We quickly realized that these data could be linked with studies of swarming rates to determine how the Africanized bees could achieve the high swarming rates we were recording.

Indeed, our studies were providing key information concerning how these bees could spread so quickly. We were in a swarm-catching frenzy, sometimes handling two or three swarms a day. We monitored about forty colonies closely, going out to each of five apiary sites daily to check for swarms. It is possible to predict, almost to the hour, when a colony will swarm, and we often arrived at the apiaries just as the familiar roar of bees began. We developed a keen intuition for locating swarms in the surrounding bushes and trees; it was the rare swarm that escaped our attention, and, once seen, it was even rarer for a swarm to escape capture.

Swarms can, in theory, be easily shaken into a hive, or into a mesh bag and then transferred to a hive; as long as the queen is caught, the workers usually will stay in the hive and begin a new colony. Swarm catching is not always that easy, however. Swarms like to settle high in trees, on limbs out of easy reach. Nevertheless, we persisted; Gard would be the "top man," climbing out on the limb where the swarm was; I would be the "bottom man," holding the bag under the swarm. In a carefully timed jump, Gard would shake the swarm from the limb, the bees would cascade into the bag, and I would pull the drawstring. The captured swarm could then be transported to one of our apiaries, where we would transfer it into its new home.

From the number of trees we were climbing and the number of swarms we were catching, it was quickly apparent that Africanized bees were maintaining an unusually high reproductive rate. This level was maintained by a combination of short generation times and multiple swarms produced during each cycle. Colonies were swarming every fifty to sixty days; many swarmed three or even four times during the eight-month dry season. Further, each colony produced a prime swarm and an average of two afterswarms during each swarming episode. Thus, a typical colony would produce about twelve swarms during the dry season. Gard estimated that one colony and its offspring would produce sixty-four swarms during this reproductive season, resulting in an astounding potential population growth rate for Africanized bees.

Real populations experience mortality, however, and the bees in our study were no exception. We did not protect colonies from the abundant ant and mammal predators that typically inhabit tropical forests, and attacks were frequent. Army ants are the most important predator in French Guiana, initiating epic battles of the social insects. These ants, whose colonies sometimes number in the millions, hang together in a loose bivouac at temporary nest sites in the forest. Columns of ants are sent out to meander through the jungle searching for prey, and they are particularly fond of other social insects whose nests contain a tasty concentration of brood. When prey such as a bee, wasp, or termite nest is discovered, the ants quickly mount an attack at the nest entrance, ganging up on the individual adult defenders and tearing them to pieces. Eventually, if the ants succeed in entering their prey's nest, they will remove the developing brood and carry it back to their bivouac site for consumption.

Mammals and birds are the other major predators in French Guiana. Of the mammals, the anteaters are particularly bothersome. These large, slow animals hide high in the trees during the day, descending at night to feed. They have an inordinate fondness for bees; they knock over hives and lick up the brood and honey, their heavy fur preventing stings from the furious bees. An anteater attacks one or two colonies a night, moving on only after all of the colonies in an apiary are destroyed. Birds are predators as well, notable because they eat the relatively large queens flying to and from the colonies on mating flights. Queen loss from such predation, or from queens failing to orient back to their nests when they go on mating flights, is another significant mortality factor for colonies.

We found that half of the colonies we followed died or abandoned their nests within seven months, and it was unusual for a colony to survive for more than a year. Nevertheless, when Gard combined the data showing high reproductive rates with the mortality data, he found that population growth rates in French Guiana resulted in a sixteen-fold annual increase in colony numbers (Figure 4). These data, consistent with observations from all over Latin America, explain why the density of Africanized bees increases so quickly after their initial arrival in a new habitat.

We still did not know how the Africanized bees could maintain their phenomenal swarming pace. We had kept some European colonies in French Guiana, and found that they would dwindle and die at the very time the Africanized bees were thriving and swarming. Even in a more suitable temperate habitat, the highest birth rate ever reported for European bees was 3.6 swarms per year, from a study in Kansas;

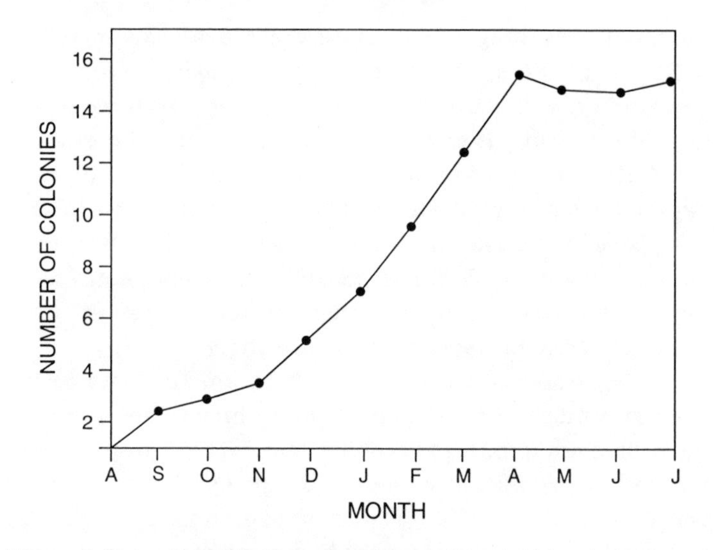

Figure 4. Rate of growth of the population of feral Africanized honey bee colonies in French Guiana, 1976–1977.

this was only 5 percent of the rate of 64 colonies per year that we found in French Guiana. We began to examine worker and colony characteristics of the two races in more detail, seeking some explanation for this reproductive variability.

The most obvious difference between temperate and tropical honey bees that is important for swarming is nest size. European bees invest a considerable amount of energy in constructing large nests and storing honey. As we have seen, nests of Africanized bees are small and store relatively little honey, so that more of a colony's energy can be invested in swarms. Further, because Africanized bee workers are smaller than European bees, less food and energy are required to rear each individual.

Another important difference is in the proportion of young to old workers. A large number of young workers issuing with swarms is advantageous because these bees have a potentially longer life span in the newly established colony, and new workers will not begin to emerge for at least three weeks after the swarm has colonized a new nest site. Almost all of the Africanized workers in a colony under the age of eight days old will swarm, whereas only 70 percent or fewer European bees in this age class issue with swarms. The large number of young workers in Africanized colonies promotes more rapid initial colony growth, and a shorter interval until the newly established colony can itself swarm.

Finally, tropical-evolved honey bees are tuned to reproduce at a much earlier point in colony growth and development than are temperate-evolved bees. That is, tropical bees respond to significantly lower "set points" of the stimuli which induce workers to rear queens and subsequently swarm. For queen rearing to begin, a number of factors have to be at appropriate levels, including colony size, degree of congestion, age distribution of workers, and queen odors (which normally inhibit workers from rearing queens). Also, abundant resources in the field seem to be desirable before colonies prepare to swarm. Africanized colonies (1) are smaller in volume and adult population than European bees when queen rearing is initiated, (2) grow to a congested state more quickly, particularly in the area where brood is being reared, and (3) show a higher proportion of young workers in colonies when swarming occurs. Nothing is known about differences in the fourth factor, queen odors. In addition, Africanized bees will swarm when resources are much less abundant in the field than the level that seems to be required for European bees.

The high swarming rate of Africanized bees has proven to be an ideal preadaptation for its colonization of the Americas. For one thing, it has allowed the bees to expand rapidly, and at high densities, by occupying an available ecological niche in the New World tropics and/or by outcompeting previously resident solitary or social bees. Also, the short intervals between successive swarmings, and the large number of swarms produced, have the effect of offsetting high colony death rates due to predation. The larger colonies characteristic of temperate-evolved bees are more likely to be attacked before swarming, both because of the longer period before they attain sufficient size to swarm and because of the attractiveness to predators of a large nest.

Unfortunately, the pronounced tendency of Africanized bees to swarm, which is highly adaptive in the wild, has had a significant impact on honey production in managed colonies. The essence of good beekeeping is to prevent swarming, thereby manipulating colonies into growing to the abnormally large sizes coincidental with honeyflows, the periods of intense nectar production in the field (Figure 5). These large colonies have surplus bees that are not needed for jobs inside the nest and can be recruited into exploiting the copious amounts of nectar available during good honeyflows. Beekeepers attempt to prevent swarming by alleviating colony congestion; they insert empty frames of comb into crowded areas of the colony and add extra boxes called supers to accommodate the storage of incoming nectar during honeyflows. Not only do these procedures discourage swarming, they also encourage colonies to grow to the size needed to collect surplus quantities of nectar.

Difficulty in preventing Africanized colonies from swarming is, in my opinion, the single most important management problem facing beekeepers working with these bees. In theory the techniques of swarm prevention used for European bees should work with Africanized colonies. But beekeepers are reluctant to approach colonies and perform any manipulations because of the aggressive nature of the bees. Ironically, the large colonies most in need of management also are the most aggressive and the most difficult to handle. In addition, Africanized bees may not be as easy to dissuade

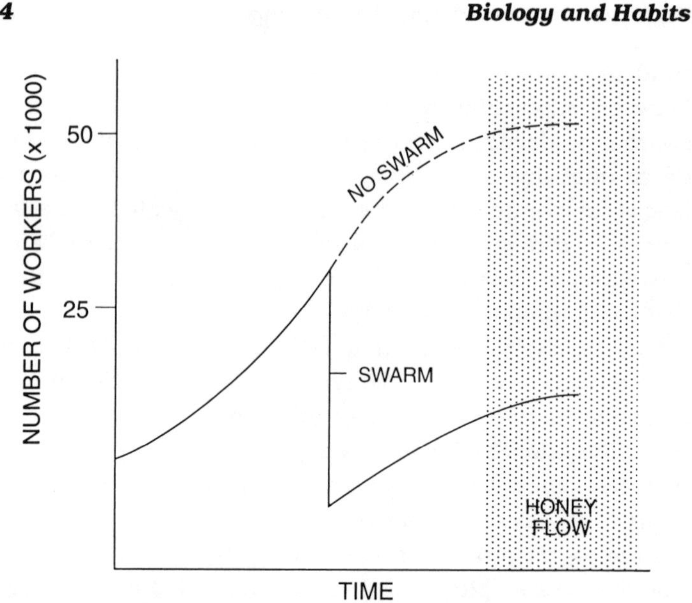

Figure 5. The impact of swarming on colony worker population relative to the timing of the honeyflow (when plants are producing the most nectar in the field).

from swarming as European bees; they frequently swarm even when appropriate antiswarming manipulations are performed. Furthermore, even the slightest manipulation of Africanized bees can result in a colony's abandoning its hive, which of course does not do much for honey production. Finally, the high swarming rate of Africanized bees has greatly increased competition for nectar from the large number of colonies now found in the wild, so that honey production in managed colonies is problematic at best.

High rates of reproductive swarming are not the only problem facing beekeepers working with Africanized bees; these bees also have a strong probability of *absconding*. That is, a colony that forms a swarm abandons its nest and reestablishes itself elsewhere. Absconding swarms differ from reproductive swarms in that no adult individuals are left behind when the swarm issues, so the original colony is no longer active. An average of 30 percent, and up to 100 percent, of colonies in an apiary abandon their hives each year. When these high absconding rates are coupled with high swarming rates, it is quite apparent why beekeeping in Latin America has suffered since the arrival of Africanized bees.

Absconding is difficult to study because, unlike reproductive swarming, there are no clear indicators of absconding events. The two types of absconding are disturbance induced, which can occur at any time during the year, and seasonal or resource induced, which in French Guiana is concentrated during the wet season. We were able to document many disturbance-induced abscondings, since we were at our apiaries almost daily, and we either encountered predators disturbing colonies or could piece together what had happened from the condition of the abandoned colonies. Because we fortunately decided to continue our measurements of seasonal absconding and our examinations of colonies during the wet season, a more definitive picture of preabsconding behavior gradually emerged.

The principal disturbance that leads to absconding is predator attack; if ants or mammals succeed in breaching the nest defenses and consuming brood and/or stored honey, the adult bees will often abandon the nest. Heavy wasp or bird predation on bees flying in and out of the nest entrance also will cause absconding, as will fire and excessive sunlight or cold. In these situations the bees abandon their nests within hours or days, leaving behind the remaining brood and honey in addition to the wax comb they have so laboriously constructed. Manipulations by beekeepers during normal colony management may also disturb the bees enough to induce absconding, creating a real dilemma for beekeepers: colonies will swarm if the beekeeper does nothing, but sometimes abscond anyway if the appropriate management is attempted.

Seasonal absconding results from a scarcity of nectar or pollen, primarily during the wet season. It is carefully orchestrated so that virtually all of the colony's resources except for the comb are taken along by the absconding swarm. Preparations begin about three weeks in advance, with diminished brood rearing by the workers. The queen may continue to lay eggs, but they are eaten by the workers; only the brood in the colony at the time is reared to adulthood. The colony waits until the last of this brood emerges before absconding, since these young workers will provide relatively long-lived individuals with which to found the new nest. Workers also engorge with honey before leaving, generally carrying all of the remaining stored honey with them.

These absconding swarms behave very differently from reproductive swarms and seem designed for long-distance dispersal. When we dissected the stomachs of bees in various types of swarms, we found that workers in absconding swarms carry almost twice the amount of honey as workers in reproductive swarms, thereby providing ample fuel for long flights. With some workers taking more than their own weight in honey, a typical swarm carries enough honey to fuel a nonstop flight of more than 90 kilometers. Furthermore, workers in absconding swarms do not scout for nest sites after clustering, but typically travel cross-country for undetermined distances before settling at an interim clustering site. From these temporary sites the swarms send out scouts to search for floral resources, presumably moving on to another area if they do not encounter flowers suitably rich in nectar. Absconding swarms may travel 160 kilometers or more before constructing a new nest, occasionally stopping to forage and refuel before proceeding through areas of poor resources until they discover a better area with abundant localized flowering.

The combination of long-distance dispersal by absconding swarms and high rates of reproductive swarming explains how Africanized bees have spread so quickly, and rapidly reached such high densities, since their introduction to Brazil in 1956. There is continual swarm movement during the year, with the dispersal-oriented absconding swarms being largely responsible for the rapid expansion in the range of Africanized bees, and the reproductive swarms producing rapid population growth once a new area has been colonized. Further, frequent swarming and absconding continue even after colony populations have reached high densities, resulting in considerable movement of bees even within local, established populations

The high frequency of absconding for tropical-evolved honey bee colonies and the long-distance dispersal of absconding swarms are adaptive responses to the intense predation and patchy, unpredictable nature of resources that are characteristic of most tropical habitats. Disturbance-induced absconding is rare in temperate-evolved bees, largely because there are few predators to induce it. Also, absconding due to any disturbance would be a poor strategy in temperate habitats, for there would not be enough time to construct a new nest and collect sufficient honey before the winter. In contrast, predators are frequent and persistent in tropical settings, and abandoning the nest to start a new one elsewhere is the preferred strategy.

Seasonal absconding is similarly favored in tropical habitats. Honey bees have two strategies for surviving a dearth season, hoarding honey and absconding. Under temperate conditions, hoarding is the only possible strategy because of the time constraint set by winter. Consequently, temperate bees rarely abandon their nest; even in tropical regions, records of absconding by European bees are unusual. We observed that European colonies maintained in the same manner as our Africanized bees dwindled and died instead of absconding under wet-season conditions in French Guiana. Absconding has proven to be a more viable strategy for tropical-evolved bees and is particularly advantageous when resources are patchy and the duration of the dearth season unpredictable. Hoarding is less desirable, because investing energy in honey storage substantially reduces the potential swarming rate and also provides an attractive source of food for predators. Under these conditions natural selection has strongly favored absconding.

The prevention of absconding in managed colonies is time-consuming and in many cases ineffective, and beekeepers have had to adapt to high levels of colony loss. Disturbance-induced abscondings are more readily prevented than seasonal abscondings. Predator attacks by ants can be reduced by ant-proofing hive stands: the legs of the stands are set in cans filled with diesel fuel to prevent ant access. This homely technique works—but continued replenishing of the fuel barrier is necessary, and formation of even the thinnest bridge from fallen debris in the cans will allow the ants to cross and attack the hives. Situating hives in the shade and providing water sources during the dry season diminishes absconding due to overheating. Infrequent hive inspections reduce the likelihood of absconding due to management disturbance, although this lack of attention may cause other problems, such as excessive swarming or increased frequency of disease.

Seasonal absconding is more difficult to deal with, partly because it is not easy to predict which colonies will abscond and which will persist. Because seasonal absconding appears to result from poor resources in the field, supplemental feeding should constitute effective prevention. Although it is a simple matter to feed a colony sugar and have the workers store it in cells, these feedings do not seem to prevent absconding—perhaps because the rate of incoming nectar may be more relevant to absconding than the amount of stored honey. Seasonal absconding remains a serious problem for keepers of Africanized bees, and it will remain so until we more fully understand the biological basis of absconding behavior.

5

Activities outside the Nest

Bees have two main reasons for leaving their nests, defense and foraging. In both situations Africanized bees provide a dramatic example of how the ecological and economic impact generated by a colony is much greater than the sum of the activities of the individual insects that make up that colony. We have a new social insect in the Americas—one which can explode in spectacular, frenzied attacks that have killed livestock and humans, and one which may be spectacularly affecting the native communities of pollinating bees and the plants they visit. Yet it is important to remember that the Africanized bee phenomenon can be broken down into the behaviors of individual bees and colonies, which are variable and are amenable to breeding and selection toward more desirable traits, at least for the colonies used by beekeepers.

The defensive behavior of Africanized bees certainly has received the most media attention, although their long-term impact on beekeeping, agriculture, and native bee communities will be largely the result of other traits. It is difficult to assess the true importance of stinging; like any behavior, stinging is variable in these bees, and colonies show a range of defensive behaviors. All of us who work with Africanized bees have been in many situations where the bees are calm, and we wonder what the media hype is all about. We also have been in situations that are frightening and potentially dangerous, even for experienced beekeepers. These bees can be enormously sensitive to the slightest disturbance and can erupt in massive, life-threatening stinging attacks. Within seconds, a single colony can muster thousands of workers to the defense of their nest. When an entire apiary is aroused, there may be many thousands of bees attacking any moving creature near their nest, from ants to people. These attacks, though rare, are spectacular, and there have been human fatalities resulting from the receipt of hundreds or even thousands of stings in a short time.

One of the best descriptions of an Africanized bee incident comes from a letter written in February 1978 by Linda Barkley, a bat researcher who encountered a feisty bee colony in a cave near La Peca, Peru:

> Having heard of a cave which reportedly housed many bats and oilbirds, we decided to investigate . . . The cave . . . seemed an ideal place for a small tent, some cooking and collecting gear, and ample room for studying bats. About two hours after we had entered and started preparation of the noon meal and coffee, we noticed that a few bees had entered and were buzzing around those closest to the entrance and stinging them. Within minutes the entrance was full of bees. We felt compelled to retreat to the darkness at the back of the cave, since it was a hard climb from the cave down to the trail

. . . One of the men did make a run for it, but received many stings in the process . . . We were unaware of the nature of the bees and had decided we must have disturbed a nest. After watching the bees for about three hours from the relatively safe darkness, we decided this was not typical bee behavior. Bees were attacking everything! We watched as little pockets of dust billowed up where the bees were flying into the ground. Repeated attempts were made to retrieve various items from the camp and check on the numbers of bees still entering the cave. The bees, in mass, would follow the person who made such an attempt but would stop when the light became dim and only a few would follow into the dark. We decided to make our escape swinging kerosene torches around our heads. Perhaps you have had a similar experience and know how helpless and out of control you can get trying to ward off hundreds of bees from your head and face and getting stung every time you did manage to locate one in your hair . . . At this point we decided to make it to a nearby house on the trail about 100 m above the cave. One of the guides met us and related the death of his horse and mule. The mule was in a field near the stream. You could see hairs sticking up from the stingers all over his body. The horse was lying in the stream in the same condition. Needless to say, it was time to leave.

Not all incidents end as well. One of the worst attacks on record occurred in Costa Rica in 1986. A botany student, Inn Siang Ooi, from the University of Miami, was hiking up a steep hillside. He climbed over a rock and encountered a large, exposed nest of Africanized bees. The bees exploded and were upon him in seconds; unable to run because the hill was too steep, Ooi evidently panicked. He climbed or fell into a crevice in the rock and became stuck, unable to flee. Rescuers were driven away by the bee attacks, and three of them were stung so badly that they collapsed. Ooi's body was retrieved after dark, when the bees had returned to their nest, and a subsequent examination revealed that he had been stung by eight thousand bees, an average of seven stings per square centimeter. This type of fatality is different from the more typical allergic reaction in which one sting can be fatal to a hypersensitive person: Ooi's death, and most other human fatalities caused by Africanized bees, was a systemic reaction resulting from massive exposure to bee venom.

How common are these attacks? Accurate numbers are difficult to come by in Latin America, but the available data suggest that human fatality rates due to massive stinging are in the same general range as snake bites and lightning casualties. The best statistics come from Venezuela, where about three hundred fifty Venezuelans are thought to have been killed by excessive stinging from Africanized bees between 1975 and 1988, or about 2.1 deaths per year per million people. The Venezuelan fatality rate was highest in 1978, when almost a hundred people were killed; by 1985, when the public was familiar with the potential danger posed by Africanized bees and beekeepers had taken appropriate precautions to reduce stinging incidents, only twenty deaths were recorded. For comparison, there are about the same number of deaths per year in Venezuela due to malaria. In the United States, about twenty people a year die from allergic reactions to honey bee stings, a rate of 0.08 death per year per million people. This is about one-fourth the rate for lightning casualties, and three times the rate for deaths due to snake bites.

Fortunately, the mass attacks I have described are exceptional incidents, and the Africanized bee colonies commonly encountered by the public are not as dangerous. Beekeepers in Latin America go to some trouble to keep the large colonies they manage away from people and livestock, and the bees encountered away from these apiary sites do not typically cause many problems.

People most commonly see bees when swarms fly overhead or settle near roads or villages—and bees in swarms generally are quite docile. Occasionally workers in a swarm will not be carrying much honey, and these "dry" swarms can be belligerent, but such a situation is unusual. Feral colonies are a second important source of contact between people and bees, but the colonies tend to be small and again are usually docile. Large feral colonies, however, can cause serious stinging problems when they are encountered. Also, stinging incidents can develop from any size colony with enough provocation, or if the colony is at a point in its cycle when bees tend to be more defensive (as when a mated queen is not present). Finally, workers foraging at flowers do not tend to sting, although some stinging may develop at concentrated, man-made sugar sources (open bags of sugar, sweet garbage, cut sugarcane, and the like). In these situations few stings are usually received, and the bees do not pursue their victims very far because they are not defending their nest.

The sting of a single Africanized bee is essentially the same as a sting from any honey bee. A honey bee sting is made up of two barbed lancets supported by hard plates and powerful muscles, connected to a sac of venom and to specialized glands which produce alarm odors (Figure 6). When a bee stings, the lancets scissor their way into the victim, and barbs anchor the sting so that it remains in the skin when the bee pulls away. The sting continues to throb for thirty to sixty seconds, injecting additional venom and giving off alarm odors that alert other bees and mark the victim for continued attack. Meanwhile, the bee that has stung soon dies—which may be some consolation to its victims. Honey bees are unusual among stinging insects in sacrificing their defenders, presumably because additional venom is injected when the sting is left in the victim. In colonies with many thousands of workers, the loss of a few during nest defense is thus balanced by having a more potent and effective sting.

The complex blend of proteins and peptides that makes up honey bee venom is similar between races. The large number of components in the venom is due to the wide variety of insect and vertebrate pests and predators which attack bee

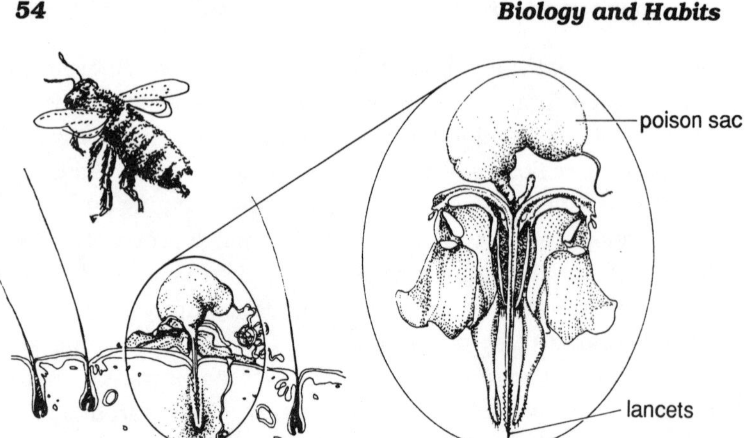

Figure 6. A honey bee sting in action. Note the large poison sac and the barbed lancets that anchor the skin in the flesh as the bee pulls away and dies.

nests; each component seems to be important in repelling different species of attackers. The principal component, and the most important to humans, is a protein called melittin. Our responses to this and other venom components take place on three levels: local, systemic, and anaphylactic. The first kind of reaction involves localized swelling, after which the affected area becomes red, itchy, and tender for a few days. A systemic reaction is more serious and may involve a rash over the entire body, wheezing, nausea, vomiting, abdominal pain, and fainting. The most serious reaction, anaphylaxis, may occur within seconds; symptoms include difficulty in breathing, confusion, vomiting, and falling blood pressure which can lead to loss of consciousness and death from circulatory and respiratory collapse. Some people are extraordinarily sensitive to bee venom and may die from a single sting. Others have or develop resistance to bee venom and can accept hundreds or even thousands of stings without a life-threatening response; one man in Africa reportedly lived after receiving

2,243 stings. Most people can tolerate only five or ten stings; systemic reactions are common with more than that number, and few people can tolerate stings numbered in the hundreds without having a serious systemic or anaphylactic reaction.

What is remarkably different about Africanized bees is that under certain circumstances colonies are capable of sudden, large-scale attacks induced by minimal disturbances. While there are certainly European bees that will sting readily, and Africanized bees that are gentle, the average Africanized colony is considerably more belligerent than most European colonies. Further, the extreme attacks that can occur with Africanized bees are almost unknown in European races.

To demonstrate the speed and intensity of nest defense that Africanized bees are capable of, we set up a stinging incident for a U.S. Department of Agriculture film crew in French Guiana. The USDA officials involved with bee publicity were troubled about a B-grade cult film entitled *The Swarm*. Due for release in the summer of 1977, the movie was of some concern because of the large number of big-name Hollywood stars appearing in it. The USDA bee group had a predicament; they did not want to produce material that exaggerated the problem to ridiculous proportions, but if they presented the Africanized bee problem as too benign, funding would not be forthcoming from Congress and the higher administrative levels of the USDA. Their goal was to produce an educational film that presented the bees as manageable, but had at least one all-out stinging attack to justify further funding.

The group asked us for help in making this film, and one morning we took the crew out to a colony that we knew from experience was going to respond with a spectacular show. We placed a small piece of leather in front of the hive, with a black circle painted on it so that the crew could film the stings against a dark background. We suited up the crew in layers of clothing, beesuits, veils, and gloves, and told them to start the cameras. Then we lightly kicked the hive. Within seconds, the air was filled with the sound of angry bees and the sweet smell of their alarm odor; within minutes, the leather cloth had received over six hundred stings. After the filming was completed, the bees followed us for almost a kilometer, and that particular colony remained difficult to approach for days.

This graphic presentation of Africanized bees at their worst illustrates a number of points. For one thing, often only a minor disturbance is needed to elicit a stinging response. We have seen that these bees can be enormously sensitive to the slightest motion, vibration, or odor, and frequently respond in group attacks. European bees, on the other hand, require a much higher level of stimulus and tend to sting as individuals, rarely rousing more than a few nestmates to join them. This difference in sensitivity is mediated by two processes. First, the threshold for stinging in individual bees is, on average, considerably lower for the Africanized bee, so less stimulation is needed before the stinging gets out of control. Africanized bees are particularly sensitive to alarm odors; one sting is very likely to lead to additional stings. Second, Africanized bees produce a greater amount of alarm odor than do European bees, thereby guiding other bees to their victim.

This incident demonstrates the persistence with which Africanized bees will pursue what they perceive as a nest attacker. While European bees rarely will pursue farther than their immediate nest vicinity, some Africanized bees will follow even a running person for up to a kilometer. Moreover, they will sting anything in sight which is moving, rapidly expanding the area they are defending and the number of victims within that area.

It is important to remember that not all Africanized bees exhibit extreme stinging behavior. Although the average colony is considerably more defensive than a European colony, there are relatively gentle Africanized bees that can be used as source material for selection and breeding. Beekeepers conduct informal selection programs by destroying particularly irritable colonies, and more rigorous breeding programs can provide queens that produce worker progeny with desirable characteristics. These queens can be substituted for queens in problem colonies, and within a few months the colony will take on the genetically based characteristics transmitted by the new queen to her workers.

Beekeepers also have a number of management options that can reduce stinging problems. The first, and most important, is to isolate apiaries to reduce the stinging threat to livestock and the public. Many countries now require that bees be kept at least 200 to 300 meters from roads, agricultural fields, and dwellings, often behind fences or barriers of vegetation. Within apiaries beekeepers also separate hives by a few meters, so that if one colony gets out of hand, neighboring colonies will not necessarily become excited as well. Beekeepers are advised to dress in many layers of clothing, to wear gloves and veils, and to use large quantities of smoke to pacify the bees prior to and during colony inspections and manipulations. They are exhorted to work the colonies quickly and infrequently.

Although these measures make for safer beekeeping, they are cumbersome and expensive to implement. Finding suitable apiary sites has been particularly vexatious, given the relatively poor state of roads in much of Latin America. Many beekeepers do not have cars or trucks, which was not a problem when household beekeeping was feasible. Colonies frequently are abandoned in beekeepers' farms or backyards simply because no vehicle is available to move them. Even after an apiary is moved, beekeepers who lack transportation cannot get to their colonies to manage them, and certainly have no way of transporting whatever honey may be produced away from the beeyards for processing and marketing. In addition, most beekeepers in Latin America are operating at levels barely above subsistence and cannot afford the expensive equipment needed for adequate protection from these bees.

Stinging problems tend to subside after Africanized bees have been at a given location for a number of years. There is a familiarization period after their arrival during which the public learns to avoid honey bee colonies; Latin American children rarely throw stones at bee hives anymore. Those beekeepers who have not adjusted to Africanized bees have left the business, and those remaining in beekeeping isolate their colonies, protect themselves while handling bees, and eliminate unusually defensive colonies. In North America, where communications and access to medical facilities are better, excessive stinging incidents will not be as common or as serious as they have been in Latin America. Nevertheless, there will always be the potential for these bees to explode in spectacular stinging incidents.

The defensive behavior of Africanized bees is another trait which, in addition to causing beekeeping problems, has ideally preadapted these bees for a feral existence in tropical America. The high rate of predator attacks in Africa has certainly favored the evolution of rapid and effective colony defense and has been adaptive in the high-predation environment of Latin America. Ironically, man has been one of the main predators on bee nests, in addition to insects and other vertebrates. Rock paintings many thousands of years old have been found that depict primitive honey-hunting techniques still in use today. Bees nesting in logs or out in the open typically are smoked first with a torch to pacify and confuse the workers, then the honey hunter reaches into the nest and removes combs containing honey, often receiving many stings in the process. Combs with brood may also be removed; the larvae and pupae are considered a particular delicacy. Certain individuals specialize in harvesting feral nests, with nest-containing trees being passed down from generation to generation as the property of a family or village. Honey hunting has been so widespread and common, especially in Africa, that it has been one of the selective pressures favoring quick and intense defensive responses by the bee workers.

The high rates of colony mortality due to predation that we found in French Guiana are similar to those in almost all tropical areas of the Americas, and the ability of these bees to defend their nests has contributed to their rapid spread and buildup. Their defensive traits, coupled with high rates of swarming and the ability to abscond when molested, have given colonies a successful battery of weapons with which to combat predators. Unfortunately, the bees do not distinguish between beekeepers and predators, and beekeepers feel the effects of defensive adaptations that have evolved over millions of years of predation in the tropics.

The other major activity that worker bees perform outside their nests is foraging. As with stinging, European and Africanized bees share similar behaviors, but subtle differences between the two in emphasis and degree result in colony-level and community-level differences that have had an enormous impact on beekeeping, and could affect native populations of other bee species. The impact and importance of a single bee is minuscule, but when their behaviors are summed into colonies—and there are millions of colonies—the reverberations of the activities of individual workers produce community-wide effects that illustrate the broad ecological and economic significance of social insects.

The two principal commodities that bees collect are nectar and pollen, although they also forage for water to drink and plant resins to insulate and strengthen their nests. Nectar, rich in sugars, is the energy source for colonies, while pollen contains most of the proteins, fats, minerals, and vitamins necessary for larval growth and adult maintenance. Plants produce nectar to attract bees because bees transfer pollen as they move between flowers, thereby effecting pollination and seed set. The worker bees also collect the excess pollen produced by flowers and bring it back to their nests. Bees require no other food; all of their nutrients are available from these floral products.

We most commonly associate nectar with honey bees, since nectar is converted to honey and stored in that familiar form. A foraging worker imbibes nectar from flowers with her long tongue, then returns to the nest carrying the nectar in a specialized sac known as the *honey stomach*. Back in the nest the nectar is transferred between workers, and enzymes are added to convert the complex sugars in nectar to simpler forms. Also, the workers add bactericidal and fungicidal agents to the nectar and regurgitate the treated nectar into cells. Workers stand over the cells and fan with their wings, evaporating water from the nectar until it is highly concentrated. Finally, the cells are capped with wax for long-term storage; the processed nectar, now in the form of honey, can be removed when the colony requires its stored sugars for energy.

Pollen is less familiar to us but is as important to bees as nectar, for it contains all of the nutrients that bees require besides sugar. Worker bees are covered with thousands of tiny plumose hairs that trap pollen grains released from flowers when bees visit, and the bees also use their mouthparts and front legs to collect the pollen more actively. Each pollen-collecting bee grooms the grains of pollen from her own body while flying, and transfers it to concave baskets on her hind legs. Then the pollen is moistened with honey and compressed into compact pellets that are carried back to the nest and packed into cells until needed. Young nurse bees consume the pollen and convert it into brood food using specialized glands in the head; this highly nutritious food is fed to larvae or other adult bees.

The collecting and processing of nectar and pollen are much more intricate than these relatively simple behaviors of individual foragers, however. For the coordination of foraging, colonies have evolved centralized recruitment mechanisms whereby scout bees find a resource and alert their nestmates to its location. Scouting workers fly over the countryside searching for nectar-producing or pollen-producing flowers. When a good food source is located, the scout returns and uses dance language to communicate the location and quality of the flowers to potential recruits waiting in the nest. These recruits then visit the flowers, return to the nest, and dance themselves if the resource is indeed of satisfactory quality. By this process of dancing and recruitment, a large foraging force can be mobilized quickly to take advantage of newly discovered resources, allowing much greater resource exploitation than would be possible from foragers working as individuals.

The dance language truly is one of the marvels of animal communication. Two types of dances are performed by returning foragers, the round dance and the waggle dance, and both provide information to naive workers in the nest concerning where and how satisfactory the floral resources are (Figure 7). The dances are performed on the comb, usually close to the nest entrance, and are observed closely by potential foragers interested in the information conveyed. In both dances the protagonists stop periodically and exchange nectar with the followers, allowing the recruits to smell the floral odors and sample the pollen scents from the balls of pollen carried on the dancers' hind legs. The profitability of food rewards is communicated by the persistence and particularly the vigor of the dance (greater abdominal vibrations and buzzing during the dance indicate a more profitable resource).

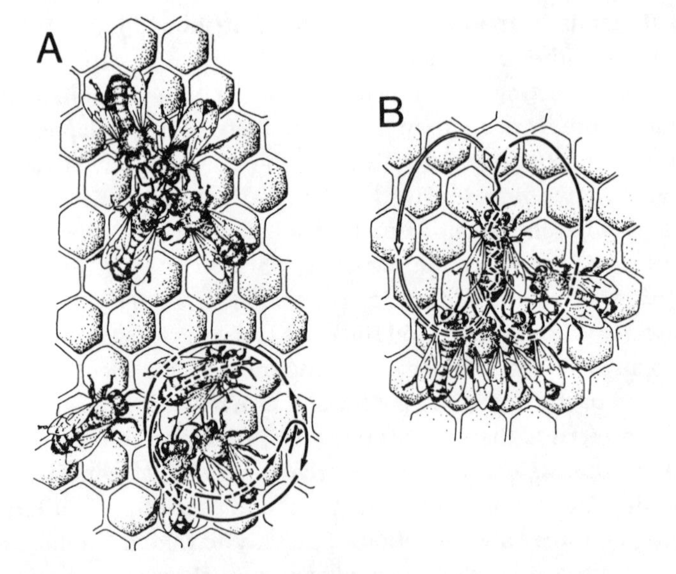

Figure 7. A. The round dance, indicating a resource close to the nest. In the top group of workers, a returning forager exchanges regurgitated nectar with potential recruits; the bottom group performs the round dance, which is being observed by recruits. B. The waggle, or figure-eight, dance, also being observed by potential recruits. From Winston 1987.

The round dance is the simpler of the two dances and is used to communicate information concerning a resource close to the nest, usually within 15 meters. The incoming worker that has discovered a nearby food source first exchanges nectar with workers inside the nest. Then she performs the round dance, repeatedly making small circles, reversing and going in the opposite direction after every one or two revolutions. This dance, although it does not convey the precise location of the resource, informs potential recruits that they should search the area immediately outside the nest for the appropriate flowers.

The waggle, or figure-eight, dance is used to communicate

information about the distance, direction, and quality of resources at distances greater than 100 meters from the nest (there are transition dances midway between round and waggle dances for resources 15 to 100 meters away). In the waggle dance, the bee runs straight ahead for a short distance, emphasizing its movements by shaking its body vigorously from side to side at a rate of about 13 to 15 times per second. The abdomen is given the most emphasis during this wagging behavior, and a buzzing sound is given off by muscular and skeletal vibrations. At the end of each straight run the bee turns in one direction and makes a semicircular turn back to the starting point, followed by another straight run and a semicircular turn in the opposite direction. As in the round dance, a retinue of workers observe closely, and food is exchanged periodically.

The remarkable aspect of this dance is the way information is translated from the abstract dance language into terms the attending workers can read and use to locate resources. For direction, the angle of the dance on the comb is transposed by the recruits into the angle away from the sun. For example, if the food is in the exact direction of the sun, the dance will be performed so that the straight run is straight up the comb. If the food is 60 degrees to the right of the sun, the straight run will be at 60 degrees to the right of vertical, and so on. Distance is communicated by the length and time of the straight run, with longer runs indicating resources that are farther away. Thus, fairly precise distance and direction information is communicated via this dance, allowing colonies to mobilize recruits rapidly to exploit desirable resources.

Both European and Africanized bees use this recruitment system, but they differ in degree; Africanized bees tend to forage more as individuals, whereas workers in European colonies use group foraging through recruitment more extensively. These differences are particularly noticeable in nectar foraging, and undoubtedly derive from differences in temperate and tropical habitats. Nectar production in temperate regions tends to be seasonally predictable, involves relatively few plant species at a time, and covers a wide geographic area. Further, individual flowers produce copious quantities of nectar with relatively high sugar concentrations, a situation referred to by beekeepers as a strong honeyflow. In contrast, tropical honeyflows tend to be weak, with flowers producing a more watery nectar; flowering is seasonally more unpredictable, geographically patchy, and involves more plant species.

The reputation of Africanized bees as copious honey producers in tropical habitats comes from their ability to collect more honey than European bees under conditions of poor, dispersed honeyflows. Under such conditions no colony will collect a great deal of honey, but Africanized colonies have been found to at least maintain their weight, and possibly show slight weight gains, whereas European colonies slowly lose weight and starve. Africanized bees are gleaners, and while they tend to collect smaller loads and return to the nest empty more often, at least some workers return to the nest with nectar during weak honeyflows. European bees tend to be "all or none" foragers, and many will not leave the nest unless the scout bees inform them that there is a reasonable probability of returning to the nest with a large nectar load. The Africanized workers also spend less time inspecting flowers and switch flowers more readily as nectar production declines than do European bees, so that they are more sensitive to subtle changes in honeyflow conditions. When flowers are secreting abundant quantities of nectar, however, European bees are more capable of using the recruitment system to mobilize a large foraging force quickly, and they collect considerably more honey than Africanized bees under those conditions.

Colonies of Africanized bees consistently show greater

numbers and higher proportions of workers foraging for pollen than European bees and may also maintain higher levels of stored pollen in their nests. By emphasizing pollen collection, Africanized bees are able to rear more brood and swarm more frequently than European bees. Also, increased pollen storage allows for some brood production during dearth periods, when European bees almost completely shut down their brood rearing. Curiously, Africanized colonies show lower recruitment rates to pollen sources, despite their greater effort and success at pollen foraging; this lower recruitment rate seems to be a consequence of a lower percentage of Africanized pollen foragers dancing upon their return to the nest. Evidently a pattern of intense group foraging is not as successful in tropical regions, where floral pollen is patchily distributed. As for nectar, the more individualistic, gleaning type of foraging style manifested in tropical-evolved bees is favored under tropical circumstances.

If the foraging behavior of Africanized bees is advantageous in tropical habitats, why have these bees been so detrimental to beekeeping? It is important to remember that the main reason Africanized bees were introduced into Brazil was their reputation as superior honey producers in the tropics, and these same bees can produce excellent honey crops in South Africa when properly managed. Also, beekeepers in South America who import European queens and try to maintain pure European apiaries have found their honey production to be considerably lower than it used to be, largely due to competition for nectar with feral Africanized nests. Indeed, feral honey bee nests were almost unknown in tropical America prior to Africanized bees, in spite of extensive beekeeping throughout the region with imported European bees; the success of feral Africanized bees provides clear evidence of how well these bees can do in Latin America.

It is not the foraging behavior of Africanized bees that is to blame for their poor beekeeping performance; in fact, they have the potential to be superior tropical honey producers. The problems for bee management are how Africanized bees utilize their foraging profits to swarm, their extreme sensitivity to subtle changes in nectar and pollen in the field (resulting in absconding), and their aggressive defense of stored resources. The same qualities that preadapted these bees for a feral existence in Latin America have prevented beekeepers from realizing the potential Africanized bees have for tropical beekeeping. Indeed, selection programs for Africanized bees that show reduced swarming, absconding, and stinging behaviors have resulted in bee stock that can do what was originally intended: adapt bees to perform well under tropical conditions. Over time, and after many generations of directed selection, the foraging behavior of Africanized bees may deliver on that promise.

The future of feral native bees in Latin America may not be as promising; Africanized bees have the potential to seriously disrupt natural pollinator communities. Bees and the plants they pollinate are a rich component of tropical forest ecosystems, and the advent of a successful competitor for the nectar and pollen that all bees require, and for the nesting sites used by other social bees, may have considerable impact on those native communities. Native bees in the tropics are a diverse and abundant group, covering a spectrum from solitary individuals to highly social species such as the stingless bees. Unusual in not having a stinger, these bees are far from defenseless; they have evolved powerful bites that often inject venom and are as effective as the stings of other bees. The stingless bees may be particularly vulnerable to competition with Africanized bees, since they nest in similar-sized cavities and forage on many of the same flowers as honey bees.

The arrival of Africanized bees has provided a huge natural experiment in which to analyze competition, but studies of extraordinary duration are necessary to demonstrate the ecological impact of an introduced species in complex environments such as tropical forests. First, many years of "before-arrival" studies are needed to establish baseline data concerning natural fluctuations of native bee populations. Then years of "after-arrival" studies may be needed to show that the potential competitor has indeed suppressed resident bee populations.

These types of studies require patience and determination. Both have been realized in the person of David Roubik, who began studying competition in 1976 as a member of the killer bee team in French Guiana. He has continued studying tropical bees in Panama, where the Smithsonian Tropical Research Institute has provided generous long-term funding.

Roubik's studies in French Guiana have provided the best evidence that competition from Africanized bees can diminish the abundance of native bees. He has censused bees periodically for thirteen years on the same one-hectare plot of acacia shrub, *Mimosa pudica,* during the wet season when little else is in bloom. Two native species of stingless bees were the major visitors to this plant prior to the arrival of Africanized bees. In 1977, about two years after the bees' arrival in French Guiana, only one in fourteen pollinators at this plot were Africanized bees (about 7 percent of the visiting bees). By 1982, 75 percent of the floral visitors were Africanized bees, and as of 1989 this figure was about 90 percent. The increase in Africanized bees was at the expense of the stingless bees, whose abundance on *Mimosa* diminished during this time. If the lower foraging numbers of stingless bees on *Mimosa* indicate a drop in their overall population, it may be in part a result of Africanized bees depleting the nectar and pollen income of stingless bees from the acacia flowers during the dearth season.

Shifts at a single floral species do not necessarily imply major changes in entire communities, however, and Roubik has initiated a much broader study in Panama to examine competitive interactions between Africanized and native bees, and the effects of shifting pollinator patterns on the plants that make up tropical forests. The Africanized bees first arrived in Panama in 1982, and few changes in the diversity or abundance of native bees or their nests have been found to date, in spite of high densities of Africanized bees. Although changes in visitation patterns at individual floral species may occur, there is considerable stability in the plant-pollinator communities found in tropical forests. Roubik estimates that community-level changes in bees may take ten years or more to become apparent. For the long-lived trees that dominate these forest systems, fluctuations in reproductive success due to changes in bee visitation patterns may take decades or even centuries to be expressed. It may be a long time indeed before we recognize the full impact of the Africanized bee on natural ecosystems.

6 The Process of Africanization

One aspect of the killer bee story that has generated considerable controversy among scientists has been the nomenclature. The word "killer" is anathema, at least when the public is listening or the media are nearby, and the taxonomic name *Apis mellifera scutellata* is too large a mouthful. The nickname "African bee" would seem appropriate, since most honey bee races are given common names based on their geographic origin; but the term does not distinguish between the bees in Africa and the bees in the Americas. It implies that the Latin American bees are still identical to their African ancestors, which may or may not be the case. Hence, "Africanized bee" has become the most commonly used name for the bees in the western hemisphere, and the term I have continued to use since our first days in French Guiana.

The controversy rages on, however, between proponents of

"African" and "Africanized"; vitriolic exchanges at scientific meetings seem far out of proportion to the topic. The passion with which each name is defended is not as silly as it appears, however. A great deal of important biology is hidden in this nomenclatural argument, with far-reaching implications for management and control programs. Two related issues are at the bottom of the dispute: (1) how similar are the bees in the Americas to the African stock that was originally introduced, and (2) how and to what extent has the Africanization of European bees occurred in Latin America? The tools of the argument have included the most modern techniques of taxonomic identification, and research designed to answer the question of "African" versus "Africanized" has made impressive contributions to a broad range of fields in the biological sciences. The results and interpretations of this research also have influenced multimillion dollar programs proposed, and in some cases implemented, to stop, control, and manage these bees.

Differentiation between the Races

The first step in addressing these broad questions is to develop techniques that can differentiate between the races, a goal that has proven to be deceptively complex. I say "deceptively" because the behavior of living Africanized and European bees at their respective nests seems easy to differentiate. We often took reporters and other visitors to see our bees in French Guiana, preparing them ahead of time via a visit to our European colonies containing queens from North America, and then taking them to the Africanized apiaries. Even the most inexperienced observer could tell them apart. The behavioral differences are striking: the European bees appear calm, slow flying, and gentle, while the Africanized bees move rapidly and are nervous, aggressive, and ready to sting. For a beekeeper these and other behavioral differences are more important than knowing whether the bees are Africanized or European; if the bees are irritable, swarm and abscond frequently, and fail to produce honey, the queen will be replaced regardless of her moniker.

The difficulty with this type of subjective taxonomy is that it does not give information about how "European" or how "African" a colony is. That is, superficial aspects of bee behavior do not differentiate hybrid bees that are a mix of European and African traits, or tell us the amount and direction of hybridization that has occurred. Yet this question is central to understanding the change from European to African characteristics. It has been necessary to deploy an arsenal of advanced scientific methodology in order to understand the identity of the feral bees and to probe the process of Africanization in managed colonies.

The history of Africanized bee systematics has paralleled developments in systematics as a whole. The early studies in the 1970s used detailed and tedious morphological measurements to compare bees of the two races, whereas more recent studies have taken advantage of biochemical techniques currently in vogue. Whatever the method, it is not a simple matter to identify different races of any organism, because there is considerable overlap in measurable characters, and even specimens representing the extreme ends of natural variability often show only subtle quantitative differences. For example, a key character used to differentiate Africanized and European bees has been the length of the forewing, which on average differs by only 0.6 millimeter between these types of bees, a difference of only 6 percent.

The first method used to distinguish Africanized and European bees was developed by Howell Daly and colleagues at the University of California. This system uses precise measurements of up to twenty-five different body parts on a single insect, which are then subjected to complex statistical analyses to derive a probability that the bee is Africanized or

European. These analyses are close to 100 percent accurate as long as at least ten bees per colony are analyzed. But it takes one person five hours to process the usual sample of ten bees, so a more rapid system has been developed by the U.S. Department of Agriculture. Called FABIS, or Fast Africanized Bee Identification System, this method begins by measuring the forewings of the sample bee, then moves on to only as many of the twenty-five body parts as are needed to yield a conclusive identification. One person can identify fifty bees per day using FABIS, which is still slow but nevertheless more cost effective than the original technique.

Analyses of South American bees clearly differentiated three groups: Africanized, European, and first-generation hybrids resulting from matings between the two. Remarkably, the feral population was almost exclusively African; it was unusual to find a wild colony that tested European or even hybrid. These findings paralleled our own behavioral and ecological studies; we too were finding virtually no feral colonies that behaved like the European type. In managed colonies, even the Europeans were becoming increasingly rare and were found only in colonies headed by queens recently imported from North America. The managed population was either African or showed some degree of hybridization between the two parent stocks. Morphologically, the hybrids were closer to European bees, but behaviorally they were more like Africanized bees, which suggested that the Africanized behaviors were genetically dominant.

The morphological analyses performed by Daly and the U.S. Department of Agriculture also revealed that the introduced bees were quite similar to the parent African stock. Many of the analyses could not differentiate between the feral bees from Africa and South America, and even where differences could be found, the Africanized bees were classified much closer to the African bees than to the Europeans. Indeed, the only major difference between the African and

South American bees was that the latter were slightly larger. Since the bees from both regions were almost identical in their swarming, absconding, and defensive behavior, it was evident that the Africanized bees were only slightly modified from their ancestors and for all practical purposes could be considered the same bee.

This conclusion has not been universally accepted. The slightly larger size of Africanized compared to African bees has been interpreted as evidence of European influence on the feral population in South America. This interpretation seems reasonable, since the European bees are larger. Still, if any increase in size of Africanized bees relative to their African predecessors has occurred, it could result from selection in South America for a larger bee or random genetic drift away from the original stock. Or such a change could be attributed to a "founder effect"—the development of a new population from a very few immigrants. In that case natural selection would work on the small part of the original gene pool that was imported, sometimes resulting in rapid divergence of the original and the founder populations.

Morphological analyses are not subtle enough to differentiate between these competing hypotheses, but the advent of sophisticated techniques in molecular biology has provided conclusive evidence supporting the concept that the feral bees are almost exclusively African, with virtually no detectable European influence.

The most definitive proof that European bees have had little or no influence on the feral population in South America has come from studies using the genetic material DNA. Differences in DNA between groups of organisms allow taxonomists to go beyond expressed and visible characters to probe the genetic code itself. The DNA in cells is made up of only four types of bases arranged in long, linear sequences, and the relative amounts and sequence of these bases are distinctive for individual species. The study of honey bee systematics has broken new ground by pushing the analyses of DNA beyond the species level in an attempt to distinguish bee races.

There are two types of DNA in cells, the DNA in the nucleus that codes for most cell functions and the DNA in the mitochondria—which are small structures in the cell that mediate energy production. The mitochondria and their associated DNA, known as mtDNA, are unusual in that they come exclusively from the mother; mitochondria from the father are not passed on after the sperm penetrates the egg. Also, evolutionary changes in the mitochondrial DNA take place only over very long time spans.

The DNA in bee mitochondria has been extracted and analyzed, and distinct differences can be detected between European and African DNA. These differences have provided a powerful tool for study of the feral bees in Latin America, and for understanding the process of Africanization, inasmuch as the presence of African DNA is possible only through an unbroken lineage of maternal descent from the original, introduced stock. The presence of European-type mitochondrial DNA can only be explained by a similarly direct descent from European stock.

What is striking about the mitochondrial DNA found in feral bees is that it is almost exclusively of the African type. These studies have used bees from locations ranging from Argentina, where the Africanized bee has been present for over twenty years, to Mexico, where the bees arrived less than five years ago. In all tropical locales, virtually no European-type mitochondrial DNA is found in any of the samples. Thus, there has been no "Europeanization" of the African bees, at least in feral colonies. Also, since mitochondrial DNA is so highly conserved, it appears likely that the DNA of the South American bees will turn out to be similar to that of their African ancestors. Indeed, current analyses of African bees indicate that the mitochondrial DNA found in feral Latin

American bees is the same type as that found in Africa, but more work is needed to confirm this finding.

Other biochemical analyses have supported the conclusion that there is little influence of European bees in the feral population. Assays of nuclear DNA have been particularly useful, since this type of DNA comes from both the queens and the drones; thus, maternal and paternal lines are represented. Reliable differences between European and Africanized bees can readily be detected, and little or no European DNA has been found in the feral population.

Glenn Hall of the University of Florida has done the most complete study of nuclear DNA, analyzing bees from Venezuela, where Africanized bees have lived for about fifteen years, and from Tapachula, Mexico, a tropical region where the bees had been present for less than two years at the time of his study. Both of these areas had large numbers of European colonies managed by beekeepers, so there certainly was opportunity for European drones to mate with the arriving Africanized queens, and for European queens that had swarmed into the wild to mate with Africanized drones. The feral Venezuelan colonies that Hall examined almost totally lacked the European nuclear DNA—even the Mexican colonies carried European genes at low frequencies, although Africanized bees had arrived only fifteen months earlier at the Mexican study site. These results are supported by other Mexican studies using morphological analysis. In one study more than 99 percent of all swarms captured near Tapachula were Africanized, while another study that dissected feral nests found fourteen of fifteen to be completely Africanized; the fifteenth was intermediate between the Africanized and European types.

Interestingly, a complex hybridization zone has been reported in northern Argentina. Bees from the more tropical regions in the northernmost parts of Argentina are of the African type, according to both mtDNA and morphological analyses, while bees from the more temperate central and southern regions are almost exclusively European. Bee samples from the zone in between show mtDNA and morphology of both types, sometimes with the mtDNA of one type associated with the morphology (that is, the nuclear genotype) of the other. Thus, while feral bees in the tropical Americas are heavily African in type, hybrid bees may occur in regions where more temperate climates allow for the survival of bees with mixed African and European characteristics.

These diverse studies, using behavioral, ecological, morphological, and genetic techniques, lead to one unambiguous conclusion: the feral honey bees in tropical Latin America have maintained their identity, little influenced by matings with European queens or drones. Further, the feral Africanized bees are similar to the original African stock that was introduced to Brazil in 1956; any subtle differences between the African and Latin American bees has come primarily from natural selection or from genetic drift away from the original African type, not from extensive hybridization with European bees. The extent of isolation has been remarkable: the Africanized bee has occupied territory previously inhabited by millions of managed European bee colonies, but from the characteristics of the feral bees found today in tropical Latin America, the European population might as well not have been there.

Isolating Mechanisms

In retrospect, much of our original research was designed to address the question of how the feral bees have maintained their African identity. Our work in French Guiana provided some of the answers, but since there were almost no European bees, we did not fully appreciate how the characteristics of Africanized bees that we were uncovering would be important in maintaining the identity of those bees in regions where managed European bees were common. French Guiana provided evidence concerning one important reason for the success of the Africanized bees: their attributes were ideally suited to tropical conditions; that is, they were extraordinarily "fit" in the tropical Americas.

The difference in fitness of European and Africanized bees can best be appreciated by the fact that, prior to the arrival of Africanized bees, feral honey bee colonies were unusual in tropical South America. European bees construct large, honey-storing colonies that swarm relatively rarely and almost never abscond, and this type of colony does not survive in tropical habitats, unless managed intensively by beekeepers. Swarms of pure European bees that establish feral nests are unlikely to survive. Similarly, queens from feral Africanized colonies that mate with drones from managed European colonies produce offspring that also would be at a disadvantage if they expressed some of the European traits. Strong selection favoring Africanized characteristics in tropical habitats would eliminate any European traits that did appear in feral colonies. European traits should predominate in more temperate areas, and hybrids of the European and Africanized bees would be more common in transition regions between temperate and tropical habitats. Such has been the case in parts of northern Argentina, although genetic analyses from that region are still preliminary.

The argument that natural selection has eliminated European characteristics from feral nest in the tropics is a compelling one, and probably is largely correct. Nevertheless, other factors may influence the predominance of Africanized bees in the wild, since feral Africanized bees have maintained their isolation even in areas where there has been a considerable amount of beekeeping with European bees (Brazil, Mexico, and parts of Venezuela). This supremacy of Africanized bees has been the case even shortly after their arrival, when managed European colonies were still numerically predominant. Some type of isolating mechanism appears to be operating; the Africanized queens and drones seem to prefer to mate with their own type rather than interbreed with their European counterparts.

To study this question, in 1978 we shifted our base of operations to Maturin, Venezuela, where the Venezuelan Ministry of Agriculture had provided us with research support and facilities in return for advice about how their beekeepers and public health officials could cope with the Africanized bees.

Our focus in Maturin was to be the mating biology of Africanized and European bees. We found that mating was much more difficult to study than the colony-level problems we had examined in French Guiana, because honey bees mate high in the air—and very quickly at that. Copulation between queens and drones occurs at congregation areas, which are discrete aerial sites where drones fly in anticipation of the arrival of virgin queens. These areas typically occupy a space 30 to 200 meters in diameter and 10 to 40 meters above the ground; many thousands of drones can be found within each area during the afternoon, when mating occurs. The drones fly lazily back and forth waiting for a virgin queen to appear, then switch to rapid pursuit when she arrives. The successful drone mounts the queen from behind while in flight, then literally explodes his semen into the queen's vagina with an audible popping sound. At this point the drone become paralyzed, drops from the queen to the ground, and dies. The entire process takes a few seconds. A queen mates with an average of ten, and up to seventeen, drones during a period of a few days, then never mates again; the sperm from these matings is stored in a special sac and lasts her for many years.

Our initial studies took place at colony entrances, partly because mating was so difficult to observe, but also because we were in the habit of observing colonies closely. It occurred to us that European and Africanized queens and drones might fly at different times of the day, which would reduce hybridization between the races. So we established an apiary with European and Africanized colonies that were each producing drones, and we also had colonies into which we put newly emerged virgin queens. We sat at the colony entrances every afternoon for four to five hours, recording the exit and entrance times of queens and drones. This was dangerous work; our apiary was located in a former mango plantation, and we had to dodge falling mangoes all afternoon—but we refused to take our eyes off the colony entrances for fear of missing a flight.

Our studies were repeated by the U.S. Department of Agriculture researchers some ten years later in western Venezuela (although without the falling mangoes), with essentially the same results: Africanized queens and drones tended to fly later in the afternoon than the European bees. The mean flight times were 5:10 and 5:14 P.M. for Africanized queens and drones respectively, whereas the mean times of European queens and drones were 4:09 and 4:57 P.M. Flights took place between 2:00 and 7:00 P.M., however, so about 70 percent of the drones of each race were flying at the same time of day. Probability values for Africanized-Africanized and European-European matings have been calculated from these and other data and show about a 60 percent chance of within-race matings. While the slight differences in the time of mating flights between European and Africanized reproductives may contribute to the isolation of the feral population, it is not sufficient to explain the predominance of Africanized traits in the wild.

Nevertheless, mating isolation could be important if Africanized drones were choosing to mate with Africanized rather than European queens, or perhaps found the European queens unattractive even in the absence of an Africanized queen. We needed to capture copulating pairs at congregation areas to determine whether such a mating preference existed, and Chip Taylor decided to devote one summer to this problem. His goal was to fly a tethered queen through a congregation area in a natural enough fashion so that a drone would copulate with her, but also so that the queen could be pulled to the ground quickly while the drone was still attached. His first device, a sort of bee leash, was demonstrated one evening in our living room. He glued a small ring onto a queen's back, tied a lightweight fishing line to it, then flew her around the room on the leash. It worked well in the living room, but never made it out to the field.

The technique Taylor settled on involved anchoring the queen in a thin plastic tube, leaving her genitalia exposed and her sting chamber pried open so that a drone could copulate with her. He devised a drone trap made of thin mesh fashioned into an inverted cone, so that drones attracted to the queen would fly up into it. These captured drones could then be pulled down to the ground and examined to determine whether they were Africanized or European.

The results of these studies were surprising; the Africanized and European drones were attracted in equal numbers to queens of either race, and *no* mating preference was apparent from the copulation data. In fact, no type of mating preference or advantage for Africanized bees has been found in any behavioral study to date. Thus, although differences in the daily timing of mating flights may have some influence, it appears that the maintenance of Africanized traits in the wild is due principally to their higher fitness in feral situations. Once the feral Africanized population builds up to high densities, matings between European drones and feral Africanized queens are unusual, because of the overwhelming number of Africanized drones at mating areas relative to the number of European drones coming from managed colonies. Further, the progeny of any European-Africanized hybrids apparently do not survive in the wild, so the feral habitat is almost entirely the realm of Africanized bees.

This conclusion is of major significance for Africanized bee control programs, since it suggests that any large-scale program to "Europeanize" the feral population through mating would be doomed to failure, at least in tropical regions. Such programs have been proposed almost since the arrival of the bees in Brazil, and just such a "genetic barrier" of European bees was attempted in Mexico to stop the spread of Africanized bees into Texas. We will look more closely at this and other control programs in Chapter 9, but our current knowledge indicates that there is little hope of influencing the feral population in tropical habitats.

Africanization of Managed Bees

The other side of the mating story is the phenomenon whereby the managed population of European bees becomes Africanized. This process of Africanization is the crux of the problem for beekeepers, since it has proven virtually impossible to maintain European bees in tropical South America, and unselected Africanized bees are virtually useless for beekeeping. The scenario has been the same throughout Latin America as Africanized bees colonize new regions: the bees arrive in low densities and initially have little effect on beekeeping. During this period the feral population of Africanized nests is low, and beekeepers notice only that an occasional colony in their apiaries acts "unusual." After two or three years the feral population reaches a critical density, and suddenly the managed bees become impossible to deal with. Stinging incidents near apiaries rise dramatically and honey production plummets; most beekeepers are forced to abandon their livelihood.

Colonies can change from European to Africanized through either mating or direct colony takeover, but mating is by far the more significant factor. The Africanized bees are not "sexier" than the Europeans; rather, the European colonies become islands in a vast sea of Africanized feral colonies and are overwhelmed by the sheer numbers of Africanized drones at mating areas. New virgin queens are produced by colonies every one to two years, and when they leave the European colonies to mate, the probability is strong that they will mate with Africanized drones. Then, when they begin laying eggs, the hybrid workers that are produced express many of the Africanized traits. The next generation of virgin queens from these colonies consists of Africanized-European hybrids, and these virgin queens again mate with Africanized drones. Thus the managed population becomes increasingly Africanized as time goes on. Even the managed apiaries become sources of the Africanized genotype unless beekeepers requeen colonies annually with imported bees of European stock.

Another type of Africanization takes place when swarms of Africanized bees land on European colonies and take them over. When this occurs, the European colonies are usually weak or temporarily without a queen. The swarm of Africanized bees settles at the colony entrance and moves into the colony, in the process killing many of the colony's defenders. The resident European queen, if present, is killed (usually by the Africanized workers), and the parasitizing Africanized queen begins laying her own eggs in the nest. The remaining European workers are subjugated to the Africanized bees, working for the new queen until they die.

Thus, although the process of Africanization results from interactions between feral and managed colonies, the direction of gene flow is almost exclusively from the feral Africanized bees to the managed European bees. Apparently the arriving Africanized bees maintain their identity in the wild owing to strong natural selection that favors their characteristics. Once the feral population has built up to high densities, which may take two or three years, the managed population quickly becomes Africanized, primarily because managed queens mate with Africanized drones, which are present in

Africanized and European worker bees in the same colony. The smaller, darker bees are Africanized. This mixed colony was created by requeening an Africanized colony with a European queen. The two types of bees overlapped in the colony for a few weeks, after the European workers began emerging but before all of the Africanized workers had died. (Photo by M. Winston.)

An external nest of Africanized bees in French Guiana, beneath a trap hive similar to those used in monitoring and control programs. Swarms usually enter these boxes and build their nests inside, but this particular colony chose to construct its nest beneath the hive. (Photo by M. Winston.)

Transferring a feral colony into a standard beehive. The colony had been nesting in the dilapidated box at bottom right. Here the combs are being removed, the contents carefully measured, and the combs strapped onto wooden bars for suspension in the hive. Once the combs and bees have been transferred to their new home, the colony can be used for research or beekeeping. (Photo by G. Otis.)

A swarm entering our apartment in French Guiana. The bees were attracted by empty hives stored inside. (Photo by M. Winston.)

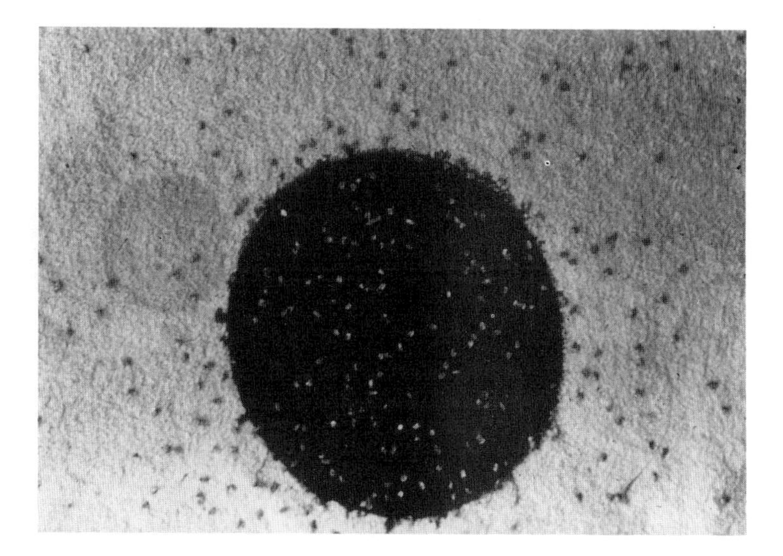

Stings from Africanized bees embedded in a leather patch. This colony was deliberately provoked to demonstrate defensive behavior for a U.S. Department of Agriculture film. The dark circle, about 15 centimeters in diameter, was used to attract bees to that particular section of the patch to film stinging in close detail. Each dot on the patch is a single sting, left behind after the rest of the bee pulled away. (Photo by M. Winston.)

An abandoned apiary in French Guiana. The colonies had been stocked with European bees, but once they became Africanized the beekeeper could no longer manage them. (Photo by O. Taylor.)

A model Venezuelan apiary, set up to demonstrate proper management techniques for Africanized bees. Note that the apiary is sheltered behind vegetation and the colonies are well separated from one another on individual hive stands. Each stand rests in cans filled with diesel oil, to prevent ant and termite access to the hives. The beekeeper, dressed in a beesuit and veil over an additional layer of clothing, is using a large smoker to pacify the bees. (Photo by G. Otis.)

Hiving a swarm. The queen is inside the hive, and the workers are orienting to her scent. Swarms usually are docile, so we are able to work without protective equipment and can hold handfuls of bees without being stung. (Photo by R. Henschel.)

large numbers. The Africanized drone population is numerically dominant for two reasons: first, there are many more colonies of feral Africanized than managed European bees, and second, those Africanized colonies produce a significantly higher proportion of drones than do European colonies of comparable size. Also, some Africanization occurs when swarms take over managed colonies. After only a few generations of matings and colony takeovers, the managed colonies are largely Africanized, although with a considerable range of hybrids represented in the managed population. At that point most beekeepers have given up; the successful ones will select hybrid and/or pure Africanized bees that can be worked with, and eventually develop stock that is manageable. The process may take ten years or more, however, and ongoing selection is necessary to maintain usable stock.

African or Africanized?

Have we resolved the question of what to call these bees? The main argument in favor of "African" is the similarity between the feral population in Latin America and bees from the Transvaal region of South Africa, where most of the first importation originated. Indeed, there is an uncanny resemblance between the African and American versions of these bees, particularly in their behavior and ecology. Bees from both habitats build almost identical nests, swarm and abscond frequently, show similar foraging behaviors, and are excessively defensive. The only detected difference is in morphological measurements of some of the South American bees that suggest a shift from the African type.

Nevertheless, the term "Africanized" has been better accepted for a number of reasons. The first is convenience; using the terms "African" and "Africanized" makes it easy to differentiate which locale is being discussed. The alternatives, "African bee in Latin America" or "neotropical African bee," are cumbersome, whereas the term "Africanized" readily distinguishes this bee as a New World tropical honey bee. Also, hybrid bees that are truly "Africanized" may be found in managed colonies and may also be present in feral colonies from habitats with a more temperate climate. Finally, the feral population will undoubtedly shift over time away from the original African bee and eventually become an easily distinguished race of its own.

II

Impact and Control

7

The Latin American Experience

Africanized bees are popularly perceived as "killers" and anomalies of the insect world. Yet in beekeeping, agriculture, and public health their influence is profound. An all-too-familiar scenario has accompanied their arrival and spread into new habitats: precipitous drops in honey production and beekeeping as well as increases in stinging incidents and fatalities of livestock and people. Over time, beekeeping rebounds to some extent and stinging incidents decrease, but the presence of Africanized bees continues to cause problems wherever they have become established. It is important to remember that prompt attention to these problems can substantially diminish their impact. The fate of beekeeping and the production of bee-pollinated crops in North America will depend on how well we learn from the Latin American experience with Africanized bees. Unfortunately, many beekeepers and government regulatory officials continue to underestimate the difficulties. Hard management decisions will have to be made when Africanized bees colonize North America, and it would be prudent to institute changes now, before the bees become widely established. If the painful experience of the Latin American beekeepers has taught us any single lesson, it is that those beekeepers who fail to anticipate and adapt to the Africanized bees will not be able to continue practicing their craft and earning their livelihood from bees.

Brazil, Venezuela, and Panama serve as useful examples to describe the impact of Africanized bees and the management techniques developed to cope with these bees in tropical habitats. All three regions were hit hard, but eventually all adopted similar management and control programs that resulted in a slow rebuilding of their beekeeping industries. These countries have been on the front lines of the Africanized bee invasion, and their experience will prove invaluable in creating lines of defense as the bees colonize North America.

Brazil

Let us begin with the First Law of Africanized Bees: the story depends on the point of view of the storyteller. To illustrate, the February 1984 issue of the *American Bee Journal,* a trade magazine read by most hobby and commercial beekeepers, contained an article entitled "Africanized Bees Now Preferred by Brazilian Beekeepers," followed by a letter to the editor that said, "It is worth remembering that the so-called 'superior' honey production by African bees has put many beekeepers out of business." These two quotes are a microcosm of the Brazilian experience with Africanized bees, and represent the two divergent interpretations of the bees' impact that have continued to the present day. Both viewpoints contain elements of truth: the Africanized bee has had a serious impact on the health and economy of Brazil, but in the long term it may prove to be the preferred bee for tropical management, particularly since beekeepers may not have any option but to use them.

The introduction of Africanized bees to Brazil was first brought to international attention in 1964, in an article published in *Bee World*, the flagship publication of the British-based International Bee Research Association. The title of this article, "The Spread of a Fierce African Bee in Brazil," gives some idea of the problems these bees were causing. The author, Paulo Nogueira-Neto, reported the rapid spread of the bees, Africanization of managed apiaries, and many stinging incidents. The article went on to describe extensive swarming and absconding, and recommended forbidding the importation of African or Africanized stock or their hybrids to new regions.

There was other evidence of serious impact: honey production decreased from 8,000 to 5,000 metric tons between 1964 and 1971, and one report suggested that over 90 percent of the beekeepers in Santa Catarina State had quit beekeeping. The Brazilians took tangible steps to respond to the developing problem: more than 23,000 virgin European queens were distributed free to beekeepers over a ten-year period in an attempt to dilute the Africanized characteristics.

Even at this early date, another perspective was emerging from Brazil, one that emphasized the superior honey-producing ability of these bees and suggested that their adverse effects had been exaggerated. This point of view was first proposed in an article by Warwick Kerr, the geneticist responsible for the original introduction, in the 1967 *South African Bee Journal*. Kerr described productivity tests in which African bees or their hybrid offspring produced two or three times as much honey as did European-derived bees. He dismissed claims of extreme defensive behavior by saying that Brazilian beekeepers were "so used to dealing with the very gentle but not very productive black bee that they regarded the new bee as impossibly vicious." Nevertheless, even Kerr went on to recommend isolation of bees from people and livestock, and the use of proper protective gear when handling Africanized bees. He finished by pointing out that penicillin kills almost a hundred persons a year in the United States, compared with six documented sting deaths in the State of São Paulo in 1966—leaving the reader to draw the obvious conclusion that penicillin was more dangerous than Africanized bees.

The divergence of opinion emerging from Brazil during the 1960s, coupled with bizarre media reports of stinging incidents, were making it increasingly difficult to evaluate the real nature and effect of Africanized bees. Consequently, the U.S. Department of Agriculture sent a team of North American scientists and beekeepers to Brazil in the early 1970s. They were to evaluate the problem, assess the likelihood that the bees would spread to North America, and recommend measures as necessary to stop the spread of the bees or minimize their impact.

The report of this committee served to crystallize the conflict between those who viewed the bees as a "serious problem" and those who saw them as a "beneficial introduction" and ignited a storm of protest in Brazil. It was based on extensive discussions with Brazilian beekeepers and officials, as well as a thorough review of the scientific research that had been performed to that time. It carefully documented the detrimental characteristics of the Africanized bee (excessive swarming, absconding, defensive behavior) and went on to describe sharp reductions in hive numbers and honey production following the arrival of the bee in new areas. The committee also recognized that selection by beekeepers and the advent of new management techniques were increasingly effective in diminishing the problems, although they made it clear that the Africanized bee continued to be deleterious to apiculture in Brazil. They concluded by proposing extensive research to better understand the bee's biology and to investigate management and control options.

Many Brazilians were not happy with this report, because the implicit conclusion was that the original introduction of African bees had not been a desirable event. This point of view was directly opposed to the developing Brazilian perspective—which was, and still is, that the Africanized bee problem has been exaggerated and that properly selected Africanized bees are superior for tropical conditions. It is perhaps not surprising that the Brazilians would adopt this stance; they introduced the bee originally and have a vested interest in the Africanized bee's living up to its good reputation. While the Brazilians have made remarkable progress in selecting and using Africanized bees, beekeeping was severely depressed in Brazil for at least fifteen years following the spread of Africanized bees. The enthusiasm with which the Brazilians embraced the bees raised expectations among many beekeepers throughout Latin America that a better bee was on the way. Unfortunately, other Latin countries do not have the extensive expertise in bee genetics and beekeeping that are found in Brazil, and the ability of Brazilian beekeepers to respond to the Africanized bees has not been duplicated elsewhere in Latin America.

Beekeeping in Brazil has bounced back from the devastation caused by the introduction of Africanized bees. Honey production has returned to previous levels and continues to grow, but much of this expansion can be attributed to extensive colonization of the Amazon Basin during the last thirty years. Vast areas of citrus and eucalyptus have been planted, providing excellent bee forage. A new road system has allowed beekeepers to practice migratory beekeeping, moving their bees long distances to take advantage of superior honeyflows. These factors would have improved honey production with pure European bees—probably to a greater extent than with Africanized bees, because the Europeans are easier to move and are better nectar collectors during the short, intense honeyflows characteristic of monocropped acreages. Finally, the price of honey in Brazil has remained high, because of lower per capita production as well as the popularity of honey as a health food.

Given the reality of the Africanized bee presence, the Brazilians have done well in selecting and improving the Africanized stock. They realized early on that attempts to maintain pure European lines would not succeed, and would not even be desirable, in the face of a large feral Africanized population. European queens have been imported to Brazil in recent years, but their colonies dwindle and die in the face of massive competition from the feral Africanized population. The Brazilians, wisely opting to select from Africanized and hybrid stock, have produced bees that are certainly more usable than those found in the wild.

Beekeeping in Brazil today prospers with selected Africanized and hybrid bees, although management procedures have been substantially modified from those used in pre-Africanized days. The early effects in Brazil were serious, perhaps more serious than the Brazilians will admit, but there is no doubt that the industry has revived. Post-Africanized beekeeping *can* work, and the Brazilians have pioneered many of the management techniques that have allowed beekeeping to continue. Unfortunately, the impact of Africanized bees was not diminished as they moved northward, and the next major beekeeping country in their path felt their full force.

Venezuela

There have been no divergent opinions concerning the Africanized bee's impact in Venezuela; it is universally recognized as disastrous. The Venezuelan beekeeping industry was in a period of expansion just prior to the bee's arrival, and the further development of beekeeping was a major agricultural priority. Surveys had revealed a rich and unexploited potential for honey production, and both hobby and commercial beekeeping of European bees was beginning to flourish.

Everything changed in 1976, when the first bees were found near Santa Eleña de Uairen in the state of Bolívar. The Africanized bees had entered Venezuela after crossing 2,000 kilometers of dense forest in the Amazon Basin. We had expected the bees' movement to be more rapid along the coast, where the habitat appeared to be superior to the tropical rain forest. Contrary to our expectations, the bees were quite capable of thriving and spreading in the jungle; they surprised us by appearing on the Venezuelan scene from the southerly direction. Survey reports indicating that most of Venezuela was good bee habitat proved all too true, and by 1981 the entire country was infested with feral Africanized nests.

The impact of these bees was dramatic. Honey production fell from about 1,300 metric tons to 78 between 1976 and 1981, and almost all of the hobby and commercial beekeepers abandoned their hives. Only two of eighteen commercial beekeepers continued to keep bees, and even those had to reduce their colony numbers by more than half. Of the hobby beekeepers, less than 10 percent maintained their apiaries. Stinging also became a major problem; close to a hundred people per year died during this period from massive numbers of stings. Ironically, the government was distributing hives, bees, and queens during this time in order to encourage Venezuelans to take up beekeeping. But the expertise of the new beekeepers was nowhere near the level needed to cope with the type of bee that was taking over their colonies.

These statistics came alive for a group of sixty American and Canadian beekeepers who visited Venezuela in 1984 to size up the situation. The tour was hosted by a U.S. Department of Agriculture research group and numerous Venezuelan beekeepers and government officials. Keith Benson, one of the beekeepers, summarized his reactions in a 1985 article published in the *American Bee Journal.* The first words of his report were, "We found that the Africanized bees were not as bad as previously represented, they were worse!" Based on the Venezuelan trip, he went on to predict that the impact of the bees in the United States would be

> of tremendous magnitude. They will not only affect the apiculture industry, but most of our other agriculture as well. They will directly affect everyone in the United States in several ways. Commercial pollination services will essentially cease to be available to the farmers in the volume now needed for production of many food and seed crops. Hobby and sideline beekeeping will virtually disappear. Commercial honey production will drop drastically . . . New laws and public liability problems will force beekeepers out of business, even if they can stand the financial loss in production . . . The invasion of the U.S. by the Africanized bee will be a major irrevocable disaster to agriculture, beekeeping and the general public. (125:188)

The situation in Venezuela today is not quite as grim, although Africanized bees still generate concern. Stinging incidents have diminished, and a better-trained generation of beekeepers has been able to cope with these insects. Today's Venezuelan beekeeper has benefited from university and government-run courses and workshops, and the formation of a national beekeeping organization has provided a much-needed forum to exchange information on management practices. Honey production has begun to increase and prices have remained high, partly because of shortages created earlier by the presence of the Africanized bees. Nevertheless, it has taken more than a decade for the first signs of rebounding, and Venezuela still has a long way to go to realize the potential of its beekeeping industry.

Panama

The arrival of the Africanized bee in Panama was particularly interesting, because Panama was well prepared for the invasion. By February 1982, when the first swarms were detected in the Darien Peninsula region, an intensive campaign of public education and bee control had already begun under the auspices of a joint United States–Panama commission. The activities of this group, (which carried the weighty name Comisión Nacional para el Control y Manejo de la Abeja Africanizada en Panama) were empowered by presidential decree. The membership included representatives from the Ministry of Agriculture, other ministries such as Health and Education, the Bank of Agricultural Development, the University of Panama, the Smithsonian Tropical Research Institute—and beekeepers. The commission's program had notably little success in diminishing the immediate impact of the Africanized bee on beekeeping, but it was highly successful in reducing stinging fatalities and in assisting the Panamanian beekeeping industry to recover from Africanization.

The apicultural statistics from Panama between 1982 and 1987 tell the familiar story of plummeting colony numbers and honey production. Dewey Caron, who has monitored the Africanized bees in Panama since their arrival, described the situation as follows. Prior to 1982, beekeeping was growing by about 10 percent a year and appeared to have a bright future as part of the Panamanian agricultural landscape. By 1987, the number of beekeepers had been reduced by half, and they were managing fewer than 40 percent of the colonies that had been producing before the Africanized bees arrived. Worse, the honey harvest had dropped to only 19 percent of its 1982 value; an industry that had previously exported honey was no longer producing enough honey even for domestic consumption. The most frightening aspect of these statistics is that the commission did an excellent job in providing information and assistance to beekeepers. The devastation occurred in spite of their efforts.

The program was considerably more successful in reducing human fatalities. Caron has estimated that on average there was only one death per year due to these bees, and there were no known fatalities prior to 1985. This level has been repeated in Costa Rica and Nicaragua, each of which has had about two deaths per year since the bees arrived.

The diminished threat to public health in Central America is not due to any change in bee behavior. Each country continues to record many hundreds of attacks on people annually; over six hundred serious attacks were recorded in Managua, the capital of Nicaragua, in 1988–1989 alone. The low number of fatalities can be attributed instead to the programs implemented throughout Central America (and now in place in most of South America) to counteract the defensive behavior of Africanized bees. The experiences of Latin American countries with Africanized bees during the last thirty-five years have been tragic, but they have resulted in an arsenal of tools with which to minimize the future impact of these bees.

The Response

The problems caused by Africanized bees in the tropics are significant, but they also are manageable. These bees seriously disrupt beekeeping and present a degree of menace to public health. Control requires major changes in bee management and also in the public's perception of bees; these changes in turn require advanced technical expertise and considerable financial resources. Beekeepers with flexibility, knowledge, and funds will continue to keep bees and prosper; those with rigid concepts, poor training, and little money will not.

Unfortunately, the resources needed to adapt to Africanized bees are beyond the reach of most beekeepers in the poorer sections of Latin America. For those who can implement them, however, the following procedures have facilitated management in Africanized regions.

Apiary locations. Backyard beekeeping—in fact, any beekeeping near people and livestock—is no longer possible with Africanized bees. Apiary locations need to be isolated and remote, at least 200 to 300 meters from roads, houses, and cultivated fields. These sites should be located behind high vegetation in order to prevent disturbed bees from spreading through the nearby countryside. The number of colonies at each site should be twenty-five or fewer, in contrast to the fifty to one hundred colonies previously maintained in single apiaries. Within each apiary colonies need to be located on individual stands, 2 or 3 meters apart, so that an aroused colony does not necessarily arouse its neighbors. The location of bee colonies has proven to be an almost insurmountable problem for beekeepers who lack vehicles or live in areas where there are few roads.

Personal protective equipment. Beekeepers used to work their European bees in shorts and sandals, with no head covering or sometimes a simple mesh veil. With Africanized bees, such attire could be fatal. Beekeepers need bee-proof clothing—a complete suit with a zippered veil that leaves no spaces for bees to enter; thick leather gloves; and high, heavy boots into which their pants can be tucked. An additional layer or two of thick clothing underneath the suit is necessary to prevent stings from penetrating to the skin, and the suit itself is often made of nylon to further reduce the penetrating ability of the stings. This outfit can be excruciatingly hot in tropical climates, but the alternative is to receive hundreds or thousands of stings while working colonies. A large smoker to pacify the bees is a required piece of equipment, and a sting kit containing injectable epinephrine is highly desirable in case someone receives a life-threatening number of stings.

Even though some of this equipment can be homemade, it may still require a financial outlay beyond the reach of many beekeepers.

Management procedures. Approaching an apiary and working colonies of Africanized bees is a hazardous occupation, and appropriate precautions are necessary. Beekeepers must put on their equipment before entering an apiary and more quickly to accomplish their objectives. Working in pairs is vital, so that one person can continuously apply smoke to a colony while the other works the bees. The ideal time is late in the day, since it is more difficult for aroused bees to find a target in the dark. Inspection and manipulation should be kept to a minimum. Colonies should be maintained at smaller sizes than with European bees, to reduce stinging and facilitate rapid management.

Beekeepers have adjusted to high levels of swarming and absconding by accepting a higher level of colony loss than experienced with European bees. To compensate, they frequently set up empty hive boxes to catch feral swarms that can replace the colonies lost from their apiaries. Of course, these swarms are predominantly Africanized and therefore must be requeened with properly selected queens.

These procedures at the beekeeper level have been useful in facilitating the performance of management tasks with Africanized bees, but have not been sufficient to deal fully with the problems caused by these insects. Governments have had to get involved by sponsoring queen selection and distribution programs, management-oriented research, extension programs, and public education. Indeed, measurable success in Africanized bee control has only been achieved in areas such as Brazil that have a sophisticated apicultural infrastructure capable of leading a multilevel response.

The selection, rearing, and distribution of usable queens is the single most important function that governments have performed. One of the world's most respected research groups in bee breeding is located in Ribeirão Prêto, Brazil, and it has been instrumental in selecting for desirable Africanized and hybrid stock. Furthermore, it has distributed large numbers of queens to replace those in beekeepers' colónies that had the more problematic feral traits. The availability of satisfactory queens has certainly helped the Brazilians to rebound from the Africanized invasion. Elsewhere in Latin America, recovery from the initial impact has been closely linked to the development of federal projects that have bred, reared, and distributed high-quality queens.

Another component of successful action plans has been agricultural extension programs that have educated beekeepers concerning the traits and management of Africanized bees. These programs may involve courses at agricultural stations, leaflets and booklets, presentations at beekeeper meetings, or traveling teams of extension agents who present short courses throughout the countryside. These extension activities emphasize how to work bees safely, methods of requeening, and (for the more advanced beekeepers) how to select and rear their own queens. Some countries supply beekeepers with equipment, and free or subsidized use of trucks to move colonies and harvest honey. Many of these programs have been funded by foreign-aid agencies, particularly in Colombia and the smaller Central American countries.

Public education has also been necessary in order to protect people and livestock from excessive stinging. Billboards, posters, leaflets, radio, and television have all been used to warn of the Africanized bee's arrival, and to discourage individuals from approaching feral or managed nests. One Panamanian poster warns people not to attack Africanized bees (Figure 8). Often notices have been posted that tell people in simple language what to do in case of stings, and provide phone numbers and/or directions to medical facilities equipped to deal with them. The distribution of anaphylactic

Figure 8. A poster widely distributed in Panama warns, "Don't attack the Africanized bees."

kits at public health units has also been significant in reducing fatalities.

Many cities have organized squads made up of trained beekeepers, members of fire departments, or other public officials, which go out each evening and destroy troublesome feral or managed colonies. These groups have been particularly helpful in keeping bees away from people in populated areas, and they may intervene in situations where a beekeeper's neighbors are being bothered by attacks. Bee SWAT teams destroy colonies by spraying kerosene, fast-acting pesticides, or other noxious agents into the nests, and may kill ten or more colonies a night in major cities.

The combination of improved beekeeper training, government action plans, and public education has been instrumental in diminishing the negative impact of Africanized bees. While no country or region has been successful in eliminating the problems caused by these bees, beekeeping has been able to continue in most areas, albeit at a reduced level. And although stinging continues to be a problem, fatalities due to massive stinging can be reduced to the status of unusual occurrences.

Bees and people have coexisted in Africa since the earliest days of human evolution, and the thriving beekeeping industry there is the best evidence that there is strong potential for Latin American beekeeping to recover from the initial impact of Africanized bees. In Africa, advanced beekeeping follows the program outlined above, using selected queens, isolated apiaries, and a management system that can accommodate swarming and absconding. Even primitive beekeeping is conducted in Africa in hollow log hives; beekeepers simply reach in and pull out comb with honey during honeyflows. The hives are small, and often suspended high in trees to keep them away from people. The colonies cannot grow excessively large owing to the continuous honey harvest. Finally, millennia of coexistence with highly defensive bees has made the public well aware of the potential dangers in disturbing colonies.

The Africanized bee problem in the tropical Americas will continue to diminish as we adjust further to the existence of a new bee type throughout the region. Indeed, the original promise of the African bee introduction may eventually be fulfilled, and perhaps we will end up with a more productive tropical honey bee. In any case, much of the story in the tropics is now history; in the more temperate regions of North America, it is just beginning. Let us turn our attention northward, to the United States and Canada, for the latest chapters of the Africanized bee saga.

8

Prognosis for North America

Predictions concerning the impact of Africanized bees in North America range from "beneficial" to "disastrous," with infinite nuances of opinion in between. There actually are optimistic beekeepers who believe that the defensive behavior of Africanized bees has been highly exaggerated and that these bees will be superior honey producers. But the bee industry also has its share of pessimists who anticipate the near-total collapse of beekeeping and vast numbers of stinging incidents. The real impact of Africanized bees in North America undoubtedly will lie somewhere between these extremes.

The reason for the lack of agreement concerning the potential impact of these bees is that many aspects of their future in North America are rather unpredictable. Two major issues limit our ability to anticipate what will happen with these bees, and many controversies have swirled around them over the last few years. The first issue involves how far north the bees will spread, and how much impact the presence of a large, managed European population will have on the feral population of Africanized bees in a temperate climate. The second issue, and probably the more problematic, concerns how beekeepers and the public will react to these bees, and how effective the response of beekeepers and government regulatory agencies will be to the problems they cause. The situation in North America may be dramatically different from that in the tropical Americas, because our scientific resources, beekeeping expertise, and agricultural extension capabilities are considerably greater than those in Latin America, and we will be working in habitats where a bee with tropical-evolved traits may be at a disadvantage. American beekeeping currently depends heavily on queen and worker bee production from the southern United States, and on migratory beekeeping that annually moves colonies on a massive scale all over the country. Also, the pollination of many crops is dependent on bee migrations. As we shall see, these aspects of American beekeeping and agriculture will be problematic when Africanized bees arrive, and the impact of the bees will depend on how flexible the industry can be in modifying current management paradigms. In this sense the situation in North America will be similar to that in the tropics; beekeepers who can adapt to new conditions will do well and prosper, while those who adhere rigidly to the old ways will not.

The question of how far north these bees will spread is obviously critical in predicting their impact. While there is general agreement that the bees will face a climatic limit, the precise location of their northernmost boundary is a much-disputed topic. Some analysts have suggested that the bees will disperse almost as far north as Canada; others propose that they will go no farther than the extreme southwestern and southeastern corners of the United States. Majority opinion has begun to coalesce around the southern third of the United States, because these tropical bees do well in subtropical regions but do not possess a number of characteristics that allow for overwintering in colder climates.

As we have seen, the ability of honey bees to endure cold winters depends on the construction of large nests containing populous colonies that store considerable quantities of honey during the summer months. The adult workers use this honey as an energy source during the winter, forming a tight cluster around it and generating body heat to keep the nest at an adequate temperature for adult survival. As the temperature drops, the winter cluster contracts and the workers generate more heat to maintain the nest temperature. There is little or no brood rearing during the late fall and early winter months, and the life span of the relatively inactive adult workers is prolonged from the 25 to 40 days typical of summer bees to 140 days or longer.

The temperate-evolved European bees presently in North America can use these wintering traits to survive in the wild in most parts of the United States and southern Canada. Feral bee colonies are surprisingly common even in areas with severe winters. For example, in upstate New York wild colonies are found at densities of about one every 2 square kilometers, even in heavily forested regions. In Kansas, where we advertised for feral colonies in order to transfer nests into the university apiaries, we located twelve nests in less than a month, eight of them on the University of Kansas campus.

Managed colonies can be found even to the northernmost borders of the Canadian provinces. Beekeepers frequently group four colonies together during the winter and wrap them with heavy fiberglass insulation. Other beekeepers move their colonies into elaborate indoor wintering facilities—large warehouses capable of holding thousands of colonies, in which temperature, light, humidity, and carbon dioxide levels are carefully monitored and regulated. In the spring the colonies are moved outdoors again.

The distribution of both feral and managed Africanized bees will be limited by their ability to survive temperate winters. There is no single trait that will prevent Africanized bees from moving into cold regions, but the combination of characteristics that make these bees "tropical" will limit their spread northward. These interrelated traits, including the small size of nests, choices of nest site, limited honey storage, short worker life spans, frequent swarming and absconding, and metabolic differences from European individuals and colonies, make winter survival difficult for Africanized bees.

Nest characteristics are important to winter survival in that many of the sites chosen by Africanized bees would not be appropriate in cold climates. For example, external nests are extremely unlikely to survive prolonged winters because they lack any insulating capability beyond the bees themselves. Also, the small nests typical of Africanized bees do not produce sufficient adults to form an adequate winter cluster, and lack storage space for honey to survive more than a few weeks of cold temperatures.

The life cycle of Africanized bee colonies presents another barrier to winter survival. Frequent swarming and absconding keep the population of nests small, and honey storage low. Absconding is particularly maladaptive in temperate regions, since colonies that abandon their nests late in the season do not have time to construct a new nest before winter. Another characteristic problematic for wintering is their short life span. Few Africanized adults live long enough to survive the winter, leaving colonies with an insufficient worker population to cluster effectively and to begin brood rearing and foraging in the spring.

Individual and colony-level differences in metabolism between Africanized and European bees also suggest that Africanized bees will be relatively poor winter survivors. At cold temperatures they show higher metabolic rates, are more active, and form looser clusters than European bees. Their metabolic rate was demonstrated in one study that measured oxygen consumption in a refrigeration chamber. Africanized bees consumed 46 percent more oxygen than European bees at temperatures close to freezing, and over 100 percent more oxygen at lower temperatures. Interestingly, the Africanized bees had metabolic rates that were 54 percent lower than European bees at a more summery temperature of 23°C, indicating that tropical bees function much more efficiently in warm temperatures than they do in cold.

A number of researchers have noted that Africanized bees do not cluster well, with individuals remaining active inside the nest at temperatures well below freezing. In addition, Africanized bees when disturbed break away from the clusters they do form, which rarely occurs with European bees. The poor clustering ability and high activity level characteristic of Africanized bees undoubtedly contribute to their poor winter survival, since more active individuals have shorter life spans and more active colonies consume stored honey more rapidly.

Studies in South America with colonies kept indoors in refrigerated chambers, and in Europe with colonies outdoors, also have shown shorter (or no) survival of Africanized bees during prolonged cold spells. In a refrigerated chamber, with temperatures close to freezing, small Africanized colonies survived an average of 45 days compared to 66 days for similar-sized European colonies. Africanized colonies imported to Poland and West Germany generally did not survive the winter even when managed; survival of feral colonies would be still less likely.

Thus, it is obvious that Africanized bees will have a northern limit to the distribution of feral nests, but predictions concerning where that limit will be are still hotly debated. The line of northernmost spread will not be clearly delineated, and a transition zone of Africanized, European, and hybrid bees will be found between the purely Africanized and European regions. The zones occupied by feral nests can be subdivided into three regions: (1) purely Africanized, in which European bees and/or hybrids are rare in the wild, (2) a transition zone, with gradually decreasing Africanized traits on a south-to-north gradient, and (3) a zone of temporary colonization by Africanized bees during the spring and summer, with the residents of purely Africanized or hybrid nests dying out in the fall and winter.

These zones will not be static. There will be seasonal changes in distribution, due to both movement of swarms and to northward movement of managed colonies by beekeepers (if transportation of colonies is permitted in post-Africanized America). Annual variations in range can also be expected, depending on the differences in rainfall and temperature.

Predicting the precise climatic limits of an introduced insect is not an easy task, because it is difficult to anticipate how a new climate and a new habitat will interact to influence a foreign species. Still, we have a considerable body of information concerning the range of Africanized bees in Africa, and their history as they have moved south toward the more temperate zones of Argentina has also provided invaluable data. This knowledge, combined with the metabolic data about winter clusters and the failure of Africanized bees to winter successfully in Poland and West Germany, allows us to predict fairly accurately the potential range of feral Africanized bees in the United States.

The earliest analysis, done in the 1970s by Chip Taylor, took various climatic measures from the range of African bees in their native habitat and superimposed that distribution on the United States. Taylor used three different factors to develop his range maps: the line where −10°C is the lowest annual temperature and the lines of the 210-day and 240-day frost-free growing seasons (Figure 9A). In Africa, these measures seemed to define well the range of the African bee at the various extremes of its distribution. While the precise boundaries of the prediction for North America varied according to the parameter chosen, the general picture indicated that Africanized bees could survive and prosper in most of southern California, parts of Arizona and New Mexico, throughout all of the Gulf States, and along the eastern coast into the Carolinas.

Taylor later modified his predictions based on the southern limits of the Africanized bees' distribution in Argentina. There the range of bees moving toward the more temperate climate of southern Argentina has been stable since about 1968, with no further southerly movement (Figure 10). The zone of pure Africanization among feral nests goes down to about latitude 30 degrees south, with the transition hybrid zone extending to about latitude 32 degrees south. Temporary colonization extends to about latitude 34 degrees south, with Africanized bees rarely found south of that line, even during the summer. Taylor found that the best environmental correlate of this distribution was a line formed by the region that experienced a 16°C mean high temperature for the coldest winter month. He superimposed this climatic measure on the United States and made a somewhat more conservative prediction concerning the bees' distribution in North America (Figure 9B).

A more recent prediction based on energetic analyses has been made by three bee researchers (Southwick, Roubik, and Williams, 1990). They drew isolines showing the number of consecutive days when the normal highest temperature is below 10°C and assumed that the Africanized bees could not survive more than 90 to 120 days at these temperatures.

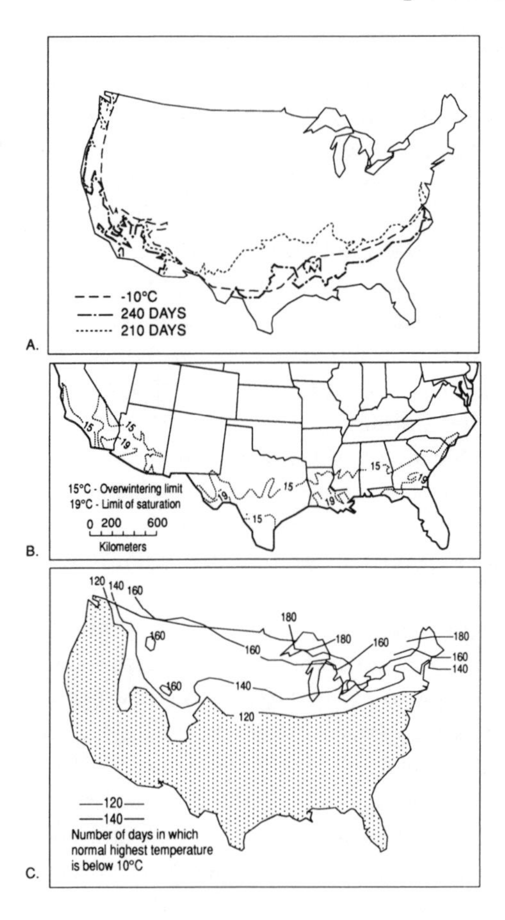

Figure 9. Several predictions of the northernmost climatic limit to the spread of Africanized honey bees in the United States. A. The 210-day and 240-day lines refer to the length of the growing season, and the −10°C line shows where this is the lowest temperature expected annually. Redrawn from Taylor 1977 (with permission of the International Bee Research Association). B. The 15° C and 19° C lines are the mean high temperature for the coldest winter months. Redrawn from Taylor and Spivak 1984 (with permission of the International Bee Research Association). C. Isolines indicate the number of days in which the normal highest temperature is below 10° C. Redrawn from Southwick, Roubik, and Williams 1990 (with permission of Pergamon Press).

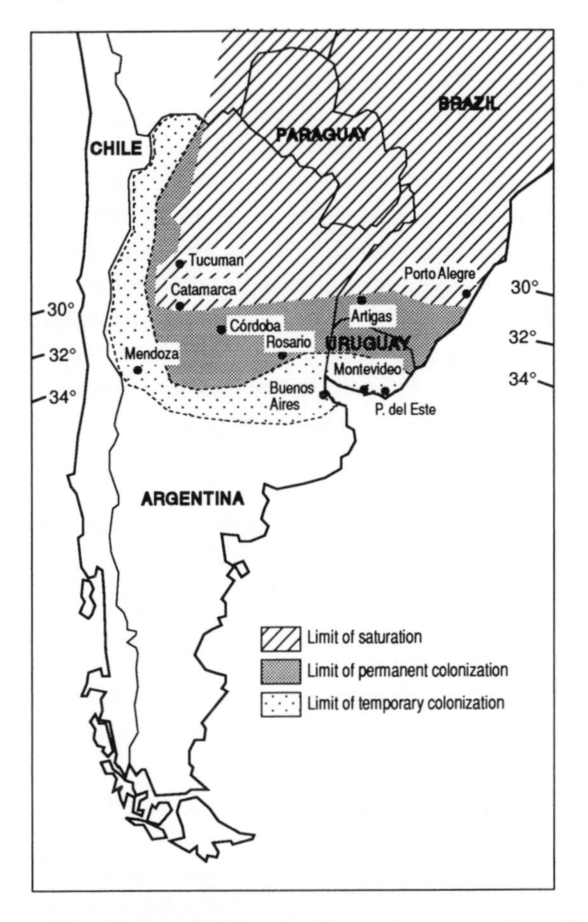

Figure 10. Distribution of Africanized honey bees in southern South America. Redrawn from Kerr, del Rio, and Barrionuevo 1982 (with permission of *American Bee Journal*).

Using the worst-case figure of 120 days, they compiled a distribution map that includes almost half of the United States (Figure 9C).

My own feeling is that the predictions of Southwick and colleagues may be a bit extreme, whereas Taylor's revised predictions appear too conservative. The most likely distribution is the compromise shown in Figure 1, with the Africanized zone extending through the southern third of the United States. The transition zone of hybrid bees will be found a few hundred kilometers north or south of this boundary, but south of this line will be the region where feral Africanized bees are common and where they will cause problems for beekeepers and the public. North of this line feral Africanized nests will be unusual, unless colonies are moved north by beekeepers. Even the most conservative predictions suggest that Africanized bees will colonize and survive in a significant portion of the United States, and it is evident that we will have to deal with these bees in at least some sections of North America.

Beekeeping in the United States may be particularly vulnerable to Africanized bees, because of the dependence of beekeepers on moving their colonies into and out of the southern states. The majority of bee colonies spend at least part of the year in the extreme south of the United States. Also, most queen and worker bees are produced in this region, then shipped to beekeepers throughout the country.

Large-scale bee movement will not be advisable after Africanized bees become widespread; the last thing U.S. beekeepers will want, or the government will permit, is beekeeper-assisted spread of these bees beyond their natural range. Even if beekeepers should wish to continue their present practice, it will take only a few stinging incidents in the North caused by bees moved up from Texas, California, or the Gulf States to shut the door on migratory beekeeping. The impact of restrictions on bee movement will be serious for beekeepers, however, and possibly catastrophic for many bee-pollinated crops; we can expect serious disagreements and political wrangles by the various agricultural and public groups that have different interests in how bees are managed.

The beekeepers most affected by Africanized bees obviously will be those who depend on migratory beekeeping.

There were about 3.3 million managed colonies of honey bees in the United States in 1989, and roughly one-third to one-half of those are moved long distances each season. The average migratory beekeeping operation is family owned and operated, and manages a few thousand colonies. Most migratory beekeepers move their colonies to the Gulf States, Georgia, Arizona, or California for the winter, to take advantage of the milder climate and earlier honey and pollen sources. In the late winter or early spring, colonies are moved from their wintering yards out to crops for pollination and to collect the concentrated floral resources available in these agricultural settings. Colonies may be moved to two or even three different crops on their way north, arriving in their northern summer locations sometime in May or June. Most migratory beekeepers in the East will winter in Florida or Georgia, midwestern beekeepers tend to move to the central Gulf States and Texas, while western beekeepers usually go to Texas, Arizona, or California. A typical migratory beekeeping operation might winter colonies in Texas, go to California to pollinate almonds and then plums, and wind up in North Dakota for the summer.

Queens need to be replaced every year or two in managed colonies, and currently about 90 percent of these replacement queens come from the deep South, again because of the early growing season and forage available in that region. In 1985 (the most recent year for which data are available), 1.2 million queens were produced for sale to beekeepers, with a value of about $7 million. In addition, most beekeepers rely on *package bees* to replace colonies that have died during the winter, or to expand their operations. These packages consist of two or three pounds of worker bees shaken from populous colonies into a wire mesh package; a queen is added, and the packages are then shipped to beekeepers and shaken into hives to start new colonies. As for queens, most package bees are produced in the South and shipped throughout the

United States in the spring. In 1986 about 900,000 packages, valued at $11 million, were produced in addition to 150,000 small colonies called *nuclei* that some beekeepers use instead of packages. A nucleus contains a queen and three to five frames of comb, brood, and bees. The nuclei produced had a value of $3.5 million in 1985. Thus, even nonmigratory beekeepers may have to deal with Africanized bees if they order queens, packages, or nuclei from the southern tier of bee-producing states.

The restrictions on migratory beekeeping and queen and package production that are likely subsequent to the arrival of Africanized bees will have a devastating effect on our beekeeping systems. As serious as the limitations on bee movement will be, these effects will be trivial compared to the impact on bee-pollinated crops. Our agricultural systems involve vast acreages of monocropped, pesticide-treated fields, in which native bees are virtually absent. Crops that require bee pollination depend almost entirely on the importation of honey bee colonies during bloom. Over two million colonies are rented by growers each year, although only one to one and a half million individual colonies are involved (many colonies are rented more than once). Beekeepers receive anywhere from $15 to $40 per colony, so that growers across the country pay beekeepers between $40 million and $50 million annually for colony rental. This figure is dwarfed by the estimated value of pollination to the crops, which in 1985 was estimated at $9.3 billion. Some of the crops involved include almonds (650,000 colonies), apples (250,000), melons (250,000), alfalfa seed (220,000), plums and prunes (145,000), avocados (100,000), and blueberries (75,000). Other crops for which growers rent colonies include cherries, vegetable seeds, pears, cucumbers, sunflowers, cranberries, and kiwis. Although about 70 percent of colony rentals occur in California, the 400,000 or so colonies resident in the state are insufficient to meet pollination needs there. As a result,

California is the state most dependent on migratory beekeeping for crop pollination.

Interestingly, on a bee-for-bee basis Africanized bees actually may be better pollinators than European bees. A higher proportion of them collect pollen, and their more rapid flight and between-flower movement could result in better pollen transfer. Still, these benefits will be overshadowed by the problems involved in moving Africanized colonies to blooming crops. Africanized bees are much more likely to abscond after a move than are Europeans, and they show larger losses in colony population when moved. The highly defensive nature of Africanized bees will make any move difficult and will generate stinging problems in fields where farm workers are present or people are passing by. But the most serious impact on pollination will come from the reduced numbers of migratory and local colonies that will be available for pollination services.

Predictions made by the U.S. Department of Agriculture in 1984 concerning the overall impact of Africanized bees on beekeeping and crop production are frightening. Estimates range from $26 million to $58 million annual losses for beekeeping alone, and an additional $93 million for crop losses due to diminished pollination (these and the following values are given in 1984 dollars). The lower beekeeping figure assumes that the bees will go no farther north than latitude 32 degrees south, while the higher figure assumes that they will spread to the 240-day frost-free line. Even the most conservative estimate of $26 million loss represents a 20 percent reduction in revenue, based on an annual total value of honey, wax, queens, and package bees of $122 million. These predictions are based on the profile of the industry as it existed in 1984 and do not take into account possible control or management measures that might diminish the impact of Africanized bees. The following are some of the main areas that will be affected.

1. Reduction or elimination of queen, package, and nucleus production from the southern United States. It is reasonable to expect that neither beekeepers nor the public will tolerate unlimited shipment of bees north from the Africanized bee zone. Even if some shipments are permitted, an expensive and time-consuming certification system will almost certainly be implemented to ensure that any bees shipped out of the zone will not be Africanized (predicted losses, $7 million to $14 million).

2. Quarantines on migratory beekeeping. Transporting bees out of Africanized bee zones will undoubtedly be reduced, and very likely prohibited, for some time after the arrival of these bees. Complete quarantines can be expected initially so a northern beekeeper who moves into an Africanized zone will not be permitted to move his colonies out. When the bees have spread through the southern states, some lateral movement may be allowed. For example, colonies might be moved from Texas to southern California and back again. Eventually a program that allows beekeepers to move colonies certified as non-Africanized might be instituted.

Migratory beekeepers who are forced to winter their colonies in the North will be faced with considerably higher operating costs, because they will have to feed copious quantities of sugar and pollen supplement in order that their colonies survive the winter and begin spring growth. Also, colony mortality of 10 to 20 percent is normal for wintering operations; replacing the lost colonies will be a major problem, given the diminished package and nucleus supply from the southern states. Finally, the elimination of migration will lower income by reducing honey yields and eliminating revenue from pollination contracts (predicted losses, $9 million to $13 million).

3. Reduced honey and beeswax production. Diminished production of hive products can be expected, due to reduced

colony numbers and a drop in per-colony productivity. The USDA study assumed that 50 to 80 percent of noncommercial colonies would go out of production, and that yields of honey and beeswax would be reduced by 30 percent in the remaining colonies. While the 30 percent figure seems valid, the assumption that all commercial colonies will remain in production seems extremely optimistic, given the experiences of commercial beekeepers in Latin America. Even these sanguine predictions show a substantial crop reduction which, in addition to its impact on beekeepers, could result in much higher prices for consumers (predicted losses $12 million to $27 million).

4. Loss of pollination fees. Quarantines on migratory beekeeping will reduce income from pollination fees for the previously migratory beekeepers. The USDA report calculated this income loss for colonies used in almond pollination and predicted a $1.4 million loss to beekeepers for this crop alone. If we make the conservative assumption that 250,000 colonies will not be available for pollination rentals and assume the low price of $20 per colony, the overall losses to beekeepers in pollination rental fees will come to $5 million annually.

5. Reductions in crop yields. Reduced colony numbers for crop pollination have the greatest potential for deleterious impact on agriculture, but these effects are almost impossible to predict with any precision. We know that a total absence of bees will result in no crop at all for many commodities, and current recommendations suggest between one and five colonies per acre, depending on how attractive the crop is. If we assume only a 1 percent reduction in crop yields over all of the United States due to inadequate pollination, crop loss will be a staggering $93 million. The USDA study predicted a $2.6 million loss to almond growers alone. Clearly, the loss to growers has the potential of dwarfing the substantial predicted loss to beekeepers.

6. Legislative and regulatory changes to beekeeping. The

October 1990 discovery of the first Africanized bee swarm in Texas (see Chapter 9) already has resulted in restrictive legislation concerning where and how bees may be kept, and the spread of the bees will trigger further regulation. Currently, hobby beekeepers keep colonies within the limits of even our largest municipalities; it is not unusual for a honey bee aficionado to keep a colony on the terrace of a high-rise apartment in the center of a major downtown area. More commonly, colonies are tucked away behind fences of backyard urban gardens, or on top of garages where the flight paths of the bees can be kept away from the neighbors. Urban beekeeping is beneficial, for the foraging bees pollinate ornamental and garden plantings. Also, urban beekeepers report excellent honey crops, because city environments have a surprising concentration of nectar-producing plants.

States and municipalities have a wide range of regulations concerning the keeping of bees in populated areas. In some cases there are no restrictions whatsoever, but more commonly bees must be kept at least 25 to 50 meters from streets and sidewalks, often behind fences. Even in some municipalities that prohibit urban beekeeping entirely, bees can be found within the city limits. The regulations are rarely enforced unless someone complains, and beekeepers have found that an annual jar of honey given to each neighbor is a great complaint reducer.

This laissez-faire situation is changing rapidly. Many states and municipalities are enacting and enforcing restrictive laws concerning apiary locations and bee movements, in anticipation of the Africanized bees' arrival. More and more, beekeepers are being prohibited from keeping bees in inhabited areas, particularly in the South, and apiary locations will become increasingly difficult to find. Liability will be another severe problem for beekeepers; it may not always be possible to obtain insurance to protect beekeepers from litigation arising from stinging incidents. Regulatory programs to certify that individual colonies are not Africanized may be implemented, at enormous cost to governments and/or to beekeepers.

7. Hobby and sideline beekeepers. Severe reductions in noncommercial beekeeping are likely, if experiences in Latin America are any indication. There are about 200,000 hobbyists in the United States defined as having 1 to 24 colonies each, and about 10,000 sideline beekeepers with 25 to 299 colonies apiece. These beekeepers will be particularly hard hit, because many of them keep bees in populated areas. The predicted 50 to 80 percent reduction in these colonies will be due to zoning regulations, insurance problems, public pressure, difficulty in managing the Africanized bees, and the intangible but important factor that the fun will be gone from a pleasant hobby.

8. Impact on the public. The most significant concern about public impact involves stinging. Although the number of fatalities will remain small, serious stinging incidents will increase, thereby creating tremendous public pressure to keep bees away from people. A more extensive treatment network for bee stings may need to be established, even though most public and private health facilities carry the appropriate medications for rapid treatment of allergic reactions to stings and other allergens. The spillover from publicized stinging incidents will extend throughout the United States, so that even northern beekeepers who do not have Africanized bees will find an unsympathetic public. Higher food prices may be another effect, if predictions about pollination problems become a reality.

Finally, tourism and recreation could be reduced in Africanized areas. Whereas the true likelihood of a traveler's being stung is low, sensational news coverage of the occasional incident could create irrational fears. This concern is not trivial; in Texas, with a $17.3 billion travel and tourism industry, even a small reduction could have deleterious consequences. One Texas official reports that he already has received calls from tourists who want to know if it is safe to venture outside major airports. A single, well-reported stinging incident at a tourist attraction can cause immense damage; imagine, for example, the repercussions at Disneyland if even one stinging incident there were publicized!

In this analysis I have used the most conservative predictions available. Even if only some of them come to pass, we will still face a serious situation once Africanized bees colonize the southern portions of the United States. Beekeeping, like most family-based agriculture, is a marginal occupation. One or two bad years can easily lead to bankruptcy. The bee industry is particularly susceptible to even minor economic perturbations, and the reduced income and increased costs that will result from Africanization may well place many commercial beekeeping operations in economic jeopardy. Furthermore, the human impact of losing a beekeeping business that may have been run by a family for generations is incalculable.

There can be no doubt that the Africanized bee is going to have a serious economic impact as it moves through North America. The potential was recognized early on and has become only too real as the bees have migrated northward. The final chapters of this story will depend on how well we cope with this bee, and whether we are successful at minimizing its potentially devastating influence.

9
Coping with the Bees

After a seemingly innocuous introduction, the drama of the Africanized honey bees burst onto the international stage almost ten years later. The plot built slowly over the next twenty-five years, filled with reports of violent attacks, exotic research efforts, and intense scientific controversy. Now we wait impatiently to discover what will happen when the bees finally arrive in our own neighborhood.

The extent and nature of their impact depend largely on our response. We are not without ideas and resources, and action plans abound at the local, state, and federal levels. All that is lacking is a consensus about what to do, and a clear line of responsibility concerning who will do what and who will pay for what. Unfortunately, the American beekeeping community has been mired in petty personal rivalries, conflicts over which organizations should speak for the industry, confusion about state and federal responsibilities, and a lack of genuine will to effect the changes in beekeeping necessary to deal with the new threat. Indeed, the most significant problems may not involve the bees themselves, but rather our lack of initiative in responding to them.

The North American response to Africanized bees thus far has involved two approaches, the first designed to keep the bees out of North America entirely, the second exploring control and management options during the initial colonization stages and the subsequent period of spread and buildup. Both approaches have been valuable in providing solid information and broad experience about Africanized bee management, and we have come to understand fairly well what will and what will not work in dealing with these bees. An enormous amount of basic and applied research has taught us more about this insect than almost any other on earth, and the combination of scientific and management data that we now possess should be sufficient to design practical and workable programs to cope with the presence of this bee in North America.

The Barrier Approach

The Africanized honey bee has been in the United States before, and in fact has been imported deliberately. Beekeepers will bring in queens from almost anywhere based on rumors of good honey production. During the second half of the nineteenth century and first part of the twentieth, beekeeping was expanding rapidly and beekeepers were searching worldwide for the bee that was best adapted to the newly settled North American continent. Advertisements in beekeeper journals offered to sell or trade queens from all over the world, including Africa, and it is possible that some queens were shipped to the United States during this time. Unfortunately, there is no reliable information concerning the frequency of importation or the fate of any African queens that might have been brought to North America.

One shipment of African queens has been well documented—an importation by scientists, not by beekeepers. In 1960 and 1961 semen from pure African drones was sent by Warwick Kerr from Brazil to the USDA honey bee genetics and breeding facility in Baton Rouge, Louisiana. Although the primary objective was to determine whether semen could be shipped long distances, the semen also was used to inseminate European queens. Hybrid bees were maintained in Louisiana for three to four years. There is no evidence that any of the Africanized genotype "escaped" into the wild—which is not surprising, since very few colonies were involved and the hybrid bees were repeatedly backcrossed with European stock. The hybrid colonies were destroyed when the implications of maintaining African bees were realized, and extensive sampling has not revealed any Africanized bees in Louisiana (or for that matter anywhere else in North America).

Africanized bees also have arrived in North America by boat, as swarms settled on Latin American ships bound for northern harbors. The first suspected swarm was reported in 1972, on a ship docked in Richmond, California, but it proved to be European. Still, at least twenty-two confirmed Africanized swarms have been identified and destroyed on boats in North American ports since 1979. They have been reported on moving boats as far offshore as 10 kilometers, where they cluster in holds, on rigging, and in containers until the boats dock. Most of these swarms have been found in Gulf Coast or California ports, but one was discovered on a boat that docked in Cleveland after coming down the Saint Lawrence Seaway. Port authorities were alerted to this problem many years ago, and most of the swarms presumably have been identified and killed. In one instance, however, a Florida beekeeper was called to collect a swarm found on a boat; rather than destroy it, he hived it in one of his colonies. Fortunately, an alert bee inspector chanced to hear a few days later that this had occurred, inspected the colony, and found that it was Africanized. The colony was, of course, destroyed.

The most extensive find of Africanized bees in North America occurred in Lost Hills, California, in 1985. A heavy-equipment operator working for an oil company observed a dead fox, apparently stung to death, and a cottontail rabbit being attacked and stung by bees coming from a fox burrow. Fortunately, the driver was in a closed, air-conditioned cab; although the bees attacked the cab, he was not stung when he dumped a load of dirt at the colony entrance. Returning a few days later, he saw bees still hovering around the burrow. He finally reported the incident to the local animal control office, and when the bees were examined, they were determined to be a mixture of Africanized, European, and hybrid bees. The nest was dug up and contained old queens cells, indicating that the colony may have swarmed. Further analysis revealed that the colony had been present for at least a year, possibly as much as two years. The nest probably originated from a swarm that had come north on an oil tanker, possibly having settled on drilling equipment that was moved from Venezuela to California.

The resulting quarantine and eradication program cost over $1 million. A task force was established immediately, and all 22,289 colonies within an area of 3,025 square kilometers around the feral nest were quarantined and could not be moved. All of the managed colonies were sampled, and as many of the feral colonies as could be located. A total of twelve colonies with some degree of Africanization were found and destroyed. All but two exhibited varying degrees of hybridization, for the presence of large numbers of European colonies in the area evidently was swamping the Africanized genes from the one feral nest. Eventually the zone was declared free of Africanized bees and the quarantine was lifted.

This exercise in Africanized bee eradication was a good dry run for future quarantines, particularly because it demonstrated the impact of a quarantine program on people. A rift quickly developed between beekeepers within the zone who wanted the quarantine lifted and those outside the zone who were afraid their colonies might be included in succeeding quarantines if Africanized bees were placed nearby. Tempers flared, and a series of impassioned meetings went on up and down California before some semblance of orderly discussion was reestablished. The already-high feelings were magnified by unfounded rumors, such as the radio report that all of the colonies within 2,200 square kilometers would be destroyed. Arizona briefly closed its borders to all bee movements originating in California, which aroused considerable indignation from migratory beekeepers who were caught with their bees in California but outside the zone. One keeper whose bees were under quarantine was caught removing boxes of honey from the zone for processing, and faced both criminal and civil charges. All of this turmoil resulted from the find of a single colony; it is hard to imagine what will occur as entire regions of the southern United States become heavily Africanized.

Indeed, the Lost Hills incident was valuable in highlighting just how difficult it will be to eradicate the bees when they do arrive in large numbers. There is a tremendous difference in scope between incidents of individual stowaway swarms arriving on boats, and thousands of swarms invading from Mexico. The preferred alternative would be to arrest the spread of Africanized bees before they reach our border, and a considerable amount of thought has gone into devising a program to meet this objective. Most experts are of the opinion that it is not possible to actually stop the Africanized bee; nevertheless, a major program to do exactly that was initiated in Mexico during the mid-1980s. This program, referred to as the Bee Regulated Zone, was coordinated through an agreement between the U.S. and Mexican governments. It was one of the most extensive and high-profile programs ever conducted to deal with any insect. Although it did not succeed in its primary objective of slowing or stopping the Africanized bee, it did provide useful information concerning how the bee is spreading, and the type of control program we may want to initiate as the bee spreads through the southern United States.

The concept of halting the Africanized bee is not new. Indeed, one of our own mandates in French Guiana was to assess the likelihood of stopping these bees, and to devise appropriate methodology if the task seemed feasible. The earliest plans were to put up some type of barrier in the narrow Isthmus of Panama region, although the nature of such a "barrier" was, and still is, vague. A number of ideas had been put forth even then, such as a high wall of gas-fueled flames along the Panama Canal or a wide radiation belt through the jungle, laid down by tactical nuclear weapons. Although these semifacetious proposals were amusing, some of the more serious proposals were equally laughable. For example, one government agency suggested that an 80-kilometer-wide swath of Central America be sprayed weekly with malathion, a pesticide highly toxic to bees and other insects, and it actually began negotiating with Central American organizations to find an acceptable region. Fortunately, this pending ecological disaster was not well received by Latin American governments, and the idea died a swift and well-deserved death.

The concept of a barrier zone began to receive more serious attention in the mid-1980s, as the Africanized bee was moving through Guatemala and into southern Mexico. Regulatory government agencies hoped that the Africanized bees would be at some disadvantage in Mexico, owing to the higher numbers of managed European colonies and the relatively subtropical climate that the bees would encounter in northern Mexico. The barrier program was worked out in meetings of the U.S. Department of Agriculture, with representatives from their Agriculture and Plant Health Inspection Service (APHIS) and Agriculture Research Service (ARS) divisions, and the Mexican Ministry of Agriculture, represented by the Secretariat of Agricultural and Water Resources (SARH). The USDA research community provided the information and scientific evaluations that encouraged the more regulatory APHIS and SARH divisions to implement the barrier program. As it turned out, the researchers' optimism was not borne out by the bees' behavior.

The objectives and plans for the Bee Regulated Zone were virtually in place by late 1985, but they were not actually implemented until almost two years later. The objective of the original plan was to install a biological barrier at the Isthmus of Tehuantepec to halt the northward progress of the Africanized honey bee. The zone was to be 225 kilometers long and 170 kilometers wide, encompassing the narrowest part of Mexico (Figure 11A). The proposed resources for the action plan included 39,000 bee colonies, 16,000 drone traps, 141,000 bait hives to catch swarms, 1,150 employees, and 220 vehicles—at a total cost of $8 million, to be shared equally between the United States and Mexico. The plan itself, couched in semimilitary terms, gave the impression that we were going to war against the Africanized bee, using sophisticated technology at the cutting edge of scientific progress. The barrier itself involved a combination of quarantines, swarm trapping, colony poisoning by pesticides, and genetic dilution of the Africanized population by European drones. More specifically, the multipronged attack was to include the following approaches.

1. *Quarantines.* Transportation of bees out of the zone was to be prohibited, to prevent the moving to the north of Africanized honey bees either purposely or inadvertently.

2. *Swarm and colony destruction.* Bait hives were to be set

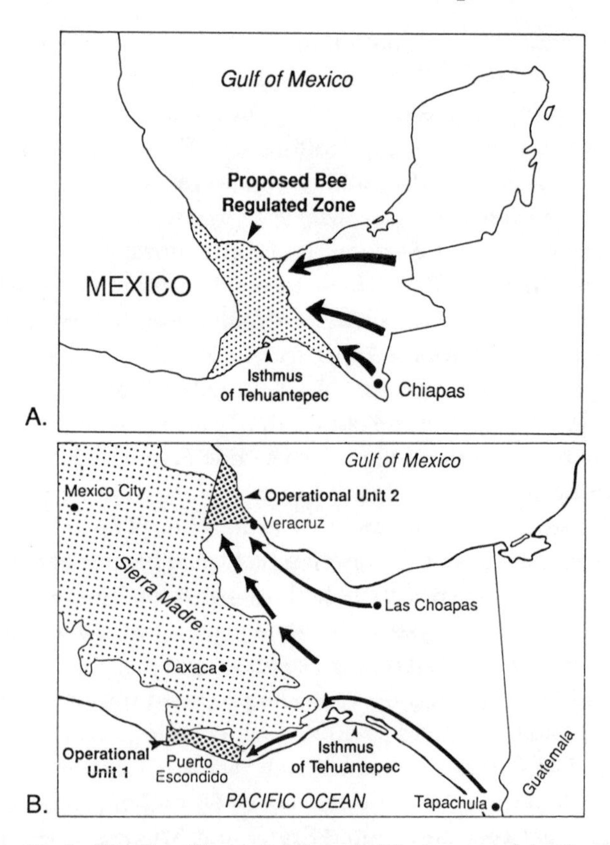

Figure 11. A. The Bee Regulated Zone in Mexico, as originally proposed. B. The Bee Regulated Zone as actually implemented, with two operational units.

throughout the region to attract swarms. The hives would be checked every week or two, and the swarms destroyed. Also, a bounty was to be offered to anyone reporting an Africanized swarm or colony; the bees would then be killed by trained personnel.

3. *Maintenance of European colonies.* All European colonies in the area were to receive queen excluders to prevent

the entry of Africanized queens in swarms. All of the European queens would be marked, and colonies would be requeened with them regularly.

4. *Drone flooding*. European colonies were to receive additional comb with drone-sized cells, to encourage the rearing of European drones. Through matings the feral Africanized genotype would thereby be swamped with European traits.

5. *Drone traps*. Traps were to be flown in congregation areas, in order to trap and kill Africanized drones. A modification of this proposal was to design the traps so that the Africanized drones would pass through a dispenser containing a slow-acting pesticide. The drones would return to their colonies, spread the insecticide through the nest, and thereby eradicate the colony.

6. *Education*. Demonstration apiaries containing Africanized bees would be set up, to show beekeepers the nature of these insects and thereby discourage intentional introduction by misguided keepers.

The proposed Bee Regulated Zone ignited a storm of controversy, with the government supporting it and nongovernment authorities lined up almost universally against it. One university scientist, quoted in the *New York Times* of 4 August 1987, called the defensive plans against the Africanized bee a doomed Maginot Line and said, "There is no way in the world we can stop them from coming into the United States." Another biologist remarked, "We're not going to stop them . . . It blows my mind to see money spent on a regulatory program that doesn't have a chance in hell of working" (*Sarasota Herald-Tribune,* 10 July 1989). The USDA officials responded by saying, "This will be the first time that humans have intervened and attempted to turn the genetic tables on the Africanized bees . . . The USDA program is on the right track, no question" (*Wall Street Journal,* 29 May 1987).

Unfortunately, the bees continued moving forward as the

rhetoric escalated, and in fact had passed far beyond the proposed zone by the time the government was ready to implement its plan. The Bee Regulated Zone eventually put in place was reduced to two regions, Operational Units 1 and 2. The first was located along the Pacific Ocean and encompassed an area 33 kilometers by 117 kilometers; the second was along the Atlantic and measured roughly 200 kilometers by 75 kilometers (Figure 11B). Both regions included some very rugged terrain, with much jungle and mountain and infrequent scattered roads. Only 150 technicians, 73,000 swarm traps, and 29,000 colonies were used (far fewer than originally proposed), at a cost of about $6.3 million during the two years the program was operational.

In addition, the objectives began to change as Africanized bees were found farther and farther north, and it became apparent that they could not be stopped. The lesser objective of slowing the bees' migration and monitoring swarm movements was substituted. By the end of the program in early 1990, the bees had moved far north of the operational units. The principal objectives changed again as the program wound down and concentrated on providing accurate data for press, legal, medical, or legislative authorities; on making more accurate predictions of what would occur after the Africanized honey bee moved into the United States; and on improving the understanding of techniques that may help to maintain European stock in Africanized areas. In short, the Bee Regulated Zone did not stop or even slow the Africanized honey bee, although it did succeed in providing information that will be useful in controlling these bees when they colonize the United States.

As might be expected, the postzone rhetoric was as inflammatory as the prezone criticism and defense. The failure of the program was blamed in part on the university scientists, who supposedly failed to support the program because they were upset at their difficulty in getting research grants from the federal government. These scientists were accused of delaying the barrier program until it had little chance of being effective. "There was a non-unified voice . . . We might have been able to launch a successful program," said one USDA official. A university staff member responded by saying, "They [the scientists] were really outgunned. The Africanized bees move too fast, survive too well. It was a formidable task. The most disturbing thing is not that they tried it, but that we aren't learning a hell of a lot from it" (*Sarasota Herald-Tribune,* 10 July 1989).

In retrospect, the Bee Regulated Zone failed to stop or slow the spread of Africanized honey bees for a number of conceptual and operational reasons.

First, the program seriously underestimated how well feral Africanized bees maintain their traits, even in areas with a considerable population of European bees. The number of European colonies maintained in the operational units, and the numbers of drones they produced, had little or no influence on the feral Africanized population. The Africanized bees were able to move through the units with impunity and build up large feral populations both inside and outside the control zones.

Second, the logistics of working in Mexico were much more difficult than expected. Quarantines simply could not be enforced, swarms were stolen from traps and transferred by peasant beekeepers to rustic log hives, and the equipment available was not sufficient for the job. Working in a tropical, developing country is not easy under any conditions, and the tasks of setting up and monitoring swarm traps, coordinating a bounty system, destroying nests, and requeening colonies overwhelmed the personnel employed by the project.

A third reason that the zone did not stop the Africanized bee is that by the time the two units became operational, the bees had already moved past them. Although the Bee Regulated Zone had been shifted north from the original isthmus site, it took only three or four months for the spreading front to move through, around, and beyond the new barrier zones. Thus, even if the project had been conceptually feasible, the operational units were simply not large enough, and were implemented much too late, to have any possibility of success.

Finally, the Bee Regulated Zone (and any other program to stop or eradicate Africanized bees) must deal with a major conceptual problem: honey bees are basically beneficial insects, so any program that targets the Africanized variety has to be benign to European bees. There are no known diseases, pesticides, pests, predators, or attractants that are selective for Africanized bees. It has even been proposed that all honey bees in Latin America be destroyed and the continent repopulated with European bees, but such ideas are in the realm of science fiction and have no basis in reality. We could, of course, spray all of the tropical Americas repeatedly with insecticide, but the cost and ecological ruin of such a program would far outweigh any problems the Africanized bee may cause in North America.

Although the Bee Regulated Zone did not achieve its goals, it was useful as a model for extension efforts to assist beekeepers in responding to Africanized bees. Program personnel trained many beekeepers in new management techniques and taught them how to identify the Africanized colonies, mark queens, and requeen colonies when needed. The program also helped beekeepers learn how to construct modern hive equipment and played an important role in disseminating information within communities.

We have learned one very important lesson from the Bee Regulated Zone: the Africanized bee is here to stay, and will soon spread through the southern United States. It cannot be stopped, eradicated, or modified in the wild, and we are left with only one option—learning to live with it. The arrival of the bee in Texas is going to usher in a new era for North American beekeeping, one that initially will be confusing, disruptive, and expensive. In the long run, we will adapt to this unusual combination of benefactor and pest. The only issue now is *how* we will adapt, and whether the steps we take will be soon enough and substantive enough to minimize the deleterious impact of the Africanized bee.

The Transition

Our response to the Africanized bee entered a new phase on 15 October 1990, when a colony of the bees was found during a routine inspection of hives near Hidalgo, Texas. By the following June fifty-nine swarms and feral colonies had been found, mostly in the Rio Grande Valley, within 15 kilometers of the border, and principally between Brownsville and Roma. The first stinging incident attributed to the northward-moving bees occurred late in May 1991, when a maintenance man riding on a lawnmower disturbed a colony nesting inside an abandoned drainage pipe and received close to twenty stings. At this writing, the southeast corner of Texas is under quarantine; bee colonies cannot be moved out of nineteen counties in which Africanized bees have been found or where their presence is suspected. Furthermore, the governor on 29 May approved an emergency appropriation of $197,000, to hire four inspectors who will enforce quarantines, monitor the location of the feral bees in southern Texas, and assist beekeepers with problems related to Africanized bees.

Beyond these immediate reactions, the transition to a post-Africanized America is by no means clear. The most serious roadblock in the way of an effective response is that each of the regulatory levels in government and industry has been passing responsibility on to another level. Indeed, the same problem exists in all insect control programs involving introduced species because of the peculiar hierarchy of federal, state, and local levels in the United States. Historically, the problem has been unusually acute in the bee industry; a particularly weak federal representation in bee regulation and an unusually divided industry send mixed messages to the legislative bodies involved.

The principal responsibility for preventing the introduction of foreign pests, and for exterminating any incipient outbreaks, rests with the U.S. Department of Agriculture, through APHIS. It is this inspection service that has been most involved in intercepting sporadic stowaway swarms, and that was the major force in the Bee Regulatory Zone. Its activities are influenced by recommendations from the ARS, and for some commodities by federal extension agents who communicate research and regulatory information to their respective commodity groups. Federal extension programs for beekeeping have been limited, although the government has moved to improve its strength in this area.

APHIS, like most politically driven federal organizations, has been quick to relinquish responsibility to state-level groups whenever controversy arises over its activities. It has been particularly quick to wash its hands of honey bee programs, because of highly vocal and mixed messages coming from different components of the bee community. For example, the inspection service withdrew from the heavily criticized Mexican program in 1990, declaring that the bees' rate of spread had been reduced and announcing that it intended to leave the forthcoming Texas invasion to the state authorities.

The habit of bowing out of controversial honey bee regulatory activities is not new to APHIS; it quickly removed itself in the 1980s from quarantine programs to prevent the spread of two honey bee pests, the tracheal and *Varroa* mites. These two foreign pests were found at various locations in the United States, and APHIS moved quickly to quarantine infected colonies. When migratory beekeepers exerted pressure to dissolve the quarantines, the service transferred mite control to state authorities, thereby facilitating a much more rapid dispersion of the mites than would otherwise have occurred. The effectiveness of the highly vocal migratory beekeepers in influencing APHIS is a source of real concern, since it is they who can potentially spread the Africanized bee far beyond its natural range.

The rationale that APHIS has used repeatedly for abandoning quarantine and eradication programs is that once a pest is established, its regulation becomes a state responsibility. Of course, each state independently interprets what is in its own interest, resulting in a hodgepodge of regulations that has little efficacy in mediating multistate and national pest problems. Furthermore, states vary tremendously in their resources and their capability to respond to introduced pests, a fact that will further promote ineffective responses to the Africanized bee.

Honey bee regulation is rendered even more complicated because the industry itself has failed to reach consensus about the Africanized honey bee, and effective lobbying of legislative bodies for funds and programs is virtually impossible without industry unity. There are two principal national beekeeping organizations, the American Beekeeping Federation and a group that splintered off from it, the American Honey Producers; these organizations generally do not coordinate their lobbying efforts, even in the rare cases when they are in agreement. In addition, each state has its own independent beekeeping organization, and these state groups do not synchronize their activities or lobbying. Thus, federal legislators and regulators hear a cacophony of opinions about how to deal with beekeeping problems; it is not surprising that they choose to waffle on regulatory programs.

Finally, there are professional research and regulatory organizations, such as the Apiary Inspectors of America and the American Association of Professional Apiculturists. Within these groups too are widely divergent opinions about virtually any honey bee topic, including the Africanized honey bee. Indeed, the battles among members of these associations have, if anything, been more divisive than the disagreements among beekeepers. Not even the professionals who are best qualified to provide leadership and guidance have been able to reach agreement about regulation and control of this new threat.

APHIS has announced that it does not intend to enact a quarantine or enforce emergency regulations in Texas when the Africanized honey bee becomes established, meaning that the state of Texas will bear responsibility for the first U.S. reaction to these bees. Recognizing this unfortunate reality, Texas has developed its own management plan. Some aspects, such as increased education, training, health care, and public information, are obvious and important steps. The core of the program involves rapid quarantines and eradication of identified Africanized colonies, as well as required annual requeening of managed colonies with European queens. The cost of enforcing these regulations is not trivial in a major beekeeping state such as Texas; the working group responsible for the plan has requested $800,000 for the first year alone. Ongoing funds have not yet been approved by the Texas legislature, and at this writing it is not clear whether the necessary financial resources will be made available.

What, then, should be done about Africanized honey bees? To my thinking, the responses and regulatory efforts needed to control these bees are obvious. We have learned from the project in Mexico and from other sources that the bees cannot be stopped, nor can we hope to influence the feral population, at least in tropical and subtropical regions. It is also evident that these bees will not spread independently beyond the southern third of the United States, and will move farther north only if assisted by migratory beekeepers. Finally, our experience with Africanized bees has made it abundantly clear that they pose a serious threat, one that should not be ignored or trivialized.

Implementation of the following actions will diminish the impact of Africanized honey bees in North America.

1. *Quarantines.* Movement of bee colonies out of regions with Africanized bees should not be permitted. Quarantines already have been established in the lower Rio Grande Valley in southeastern Texas. As the bees spread, the quarantine zone will grow; eventually unregulated north-south movement of bees will cease entirely. Lateral movement among southern states that have been colonized by Africanized bees could be permitted, however.

The seasonal need for pollinating colonies in California can continue to be met by northern bees until Africanized bees appear in that state. At that point managed colonies could be moved from any of the infested southern states into and back out of California, so that California agriculture will experience only minor disruptions. Eventually, the movement of bee colonies out of Africanized zones might be permitted if the queens were certified as European, but a major effort would be required to inspect individual colonies and might not be worth the expense.

2. *Mandatory requeening.* All managed colonies in Africanized zones should be requeened annually with marked, certified European queens. Beekeepers thereby can be assured that their apiaries will be almost completely free of Africanized influence. Besides the obvious advantage of allowing beekeepers to work with tractable bees, such a requeening program may have large-scale benefits in reducing beekeepers' liability, since those who requeen can demonstrate to the courts that they have taken reasonable precautions against Africanized honey bees. Also, requeening will make it much easier for beekeepers to obtain insurance protection.

Some level of inspection by state authorities will be necessary to enforce such a regulation, but this type of monitoring is already in place for bee diseases; it would not be difficult to expand into random queen checks. Beekeepers also could be required to furnish evidence that they have purchased the appropriate number of certified European queens for the colonies they possess. If migratory beekeeping is permitted, inspections may have to include each colony, which could be prohibitively expensive. Even so, inspection could prove to be economical, particularly if rental prices increase for the colonies used to pollinate crops.

The concept of mandatory requeening with European queens is central to Africanized bee control, but it raises the question of where these certified queens will come from. Most queens today are reared in the South, where they will soon begin mating with feral Africanized drones. Theoretically, local mating areas could be saturated with European drones, and a level of European-European matings of 90 percent or higher might be achieved. Since queens mate with up to seventeen drones, many if not most of the queens could produce some hybrid progeny, which might not be acceptable to northern states concerned with keeping their managed colonies free of Africanized bees. The progeny of each queen would have to be certified as European, a process that will vastly increase the time and expense involved in rearing a queen.

There is a much simpler way to produce certified European queens. They can be reared in parts of the world that do not have Africanized bees—on islands or in the northern United States and southern Canada. Already many queens are produced in Hawaii, New Zealand, and Australia. There is no reason why these queens cannot be selected for North American conditions, reared in large quantities, and shipped to North America. Indeed, all of these offshore locales have vibrant queen rearing industries that currently ship over a hundred thousand queens annually to the mainland United States and Canada. This industry could easily be expanded.

It might be simpler to select and rear queens suited to North America right here, in the regions where they will be used. These queens would have to be reared in late spring and during the summer, but the timing of requeening can certainly be changed from the current spring system that is dependent on southern queens to a summer system that utilizes northern-reared queens. Indeed, the Canadians have already accomplished this transition very successfully.

Changing queen supplies to offshore and northern sources and changing the timing of annual management procedures are perfect examples of how flexibility can be our strongest weapon against Africanized bees. Substantial alterations are involved, particularly for southern queen breeders and package bee producers, who will have to move their operations north. Beekeeping, like all agriculture, is as much a life style as a business, and adjustments to that life style can be daunting to families who have lived and worked at one location for generations and who are used to performing their tasks in specific ways. Nevertheless, if the options are to change or to go bankrupt, there is not much question which choice will prevail.

3. *Hobby beekeeping.* A higher level of beekeeper training will be necessary than is currently practiced, especially for hobby beekeepers who operate within city limits. Courses in beekeeping and some form of licensing should be mandatory. Workshops, symposia, audio and video tapes, and printed information should be introduced and disseminated widely to all beekeepers. Apiary locations will need to be inspected and approved for safety, and individual colonies checked annually for the presence of certified European queens.

These restrictive procedures will be necessary to protect the public, especially because even one incident might be sufficient to ban beekeeping entirely. We of the beekeeping community are going to have to prove to the public that we are operating in a safe, responsible manner, and that the beneficial aspects of properly managed honey bees are an important component of our society. Fortunately, most beekeepers are as social as their colonies and enjoy getting together to talk about bees. The organizational structure already found in most local and state communities will facilitate communication between regulatory personnel and the beekeepers.

4. *Public health and education.* The final aspect of an optimum Africanized bee program involves public health. Education is particularly important, since awareness of the potential danger from Africanized colonies has proven to be a real deterrent to getting close to feral or managed colonies. Additional training of public health officials is warranted, and wider dissemination of sting kits and information about how to use them would be advisable. We have seen that health and education programs have been successful in Latin America in decreasing the number of human fatalities, and we should be able to accomplish the same in North America.

Thus, protection from the Africanized bee is not as impossible as it might seem. Certainly the bees will have an impact: some commercial beekeepers will lose their businesses; many hobby beekeepers will decide to give up beekeeping; and it will not be easy to provide sufficient colonies to meet pollination needs. Stinging incidents will increase, and there will be some human fatalities. Nevertheless, the simple programs proposed here would go a long way toward minimizing these effects. While they may seem restrictive, they are preferable to the alternative: a very serious impact indeed.

The most important components of an Africanized bee response will be learning to adapt to the new situation and developing a consensus about post-Africanized beekeeping. Researchers, beekeeping organizations, and individual beekeepers need to cooperate if we are to achieve an orderly transition to new beekeeping paradigms. Can this be accomplished? The question has been answered resoundingly in the affirmative by Canada, a country that has successfully changed its beekeeping in the last ten years to an extent equal or greater than will be necessary in the United States. Foresight, flexibility, and cooperation have produced a new order in Canadian beekeeping, and have demonstrated that real opportunities exist for those who can keep up with the changing nature of beekeeping in North America.

The Canadian Response

Canada has met the threat from Africanized bees in a typically Canadian way: orderly change based on a slowly developing consensus. The Canadian beekeeping community was seriously threatened by these bees, because roughly half of Canadian bees and queens were imported from the southern United States.

Most Canadian beekeepers used to base their bee management on the importation of package bees from the United States each April. This system was designed to take advantage of the short but intensive northern season; flowering may last only a few weeks, but during this time Canadian crops and wildflowers produce some of the most copious honeyflows in the world. Winters are long and cold, however, so that many beekeepers removed all of the honey from their colonies before fall and killed the colonies rather than overwinter them. To fuel this system, Canada imported over 300,000 packages and queens each spring from the United States to start up the colonies again. The packaged bees worked well, but in the late 1970s Canadians began to realize that this importation system would be threatened by Africanized bees. The beekeeping community began to investigate alternatives, with the objective of making Canada self-sufficient in its need for bees and queens.

The effort was coordinated by a tightly knit community of researchers, government personnel, and beekeepers. The research and extension components of the industry were represented by the Canadian Association of Professional Apiculturists, a group that has a comfortable working relationship with the provincial beekeeping organizations, the national Canadian Honey Council, and Agriculture Canada, the federal ministry responsible for beekeeping research and regulation. These groups worked together and gradually developed a program. Research was proposed and implemented in the three areas of overwintering, queen rearing, and package bee production. These programs, and the accompanying regulations, were not without controversy. But the close communication among the groups involved, as well as the majority support of the beekeepers themselves, resulted in achievement of the self-sufficiency objective despite the objections of beekeepers who wanted to maintain package bee–oriented management systems.

The first step in this program was to generate improved technology to overwinter colonies. Two parallel methods were developed: one involved insulating colonies outdoors and the second overwintered colonies indoors in climate-controlled buildings. Both techniques proved successful, even in the coldest parts of Canada. More important, both turned out to be economically better than the package bee systems. Although overwintering involved more labor, the annual per-colony income was higher than for colonies started from packages.

There was, and still is, a need for *some* packages each spring, to replace colonies that have died during the winter. Consequently, the second component of the program involved determining the feasibility of producing package bees in Canada for spring sales. I was closely involved with this research, partially funded by the Science Council of British Columbia and with the cooperation of numerous commercial beekeepers all over Canada. We determined that package bees could be produced in southwestern British Columbia for sale in April, when beekeepers need them.

The most crucial technique for boosting bee populations to meet these April deadlines was feeding; less than one dollar's worth of extra pollen and sugar supplements fed in the fall and early spring to each colony stimulated the production of enough extra worker bees to yield additional profits of ten to twenty dollars per colony from bee sales. Interestingly, it turned out that beekeepers producing packages could make significantly higher incomes than those using colonies for honey production alone. Consequently, an active industry producing package bees has developed in British Columbia, the income from which is approaching $1 million annually. This new industry is one example of how problems such as Africanized bees can create opportunities also, for those willing to experiment a bit.

The third component of Canadian self-sufficiency is queen production. Unfortunately, it is not possible to rear queens in the early spring in any part of Canada, but this obstacle has been circumvented by changes in the timing of annual management procedures. Now most Canadian beekeepers requeen in late spring or during the summer, a cycle that has worked well throughout Canada. Another thriving industry has burgeoned, rearing queens and providing selected queen stock that has proven to be well suited for Canadian conditions. This stock resulted from government-sponsored programs across Canada aimed at producing better Canadian queens; an important side effect was the enhancement of Canadian expertise in queen selection and rearing. Today, many of those involved in the programs have gone on to start their own queen rearing businesses, another example of new opportunities drawn from the changing reality of North American beekeeping. Further, the Canadian queen rearing industry is starting to export queens, which will be helpful to American beekeepers looking for sources of non-Africanized queens.

By the mid-1980s the Canadian beekeeping community

was gradually converting to systems involving increased overwintering, British Columbia package production, and midseason requeening; the unexpected finding of parasitic mites in the United States accelerated the process. In 1987 Canada closed its borders to all bee importations from the United States, forcing Canadian beekeepers to become virtually self-sufficient much more quickly than expected. Fortunately, the necessary research had essentially been completed by that time, and government aid assisted in dealing with the effects of the abrupt border closure. Today Canada imports only a small number of queens and packages from New Zealand and Australia.

The magnitude of this change was enormous—over half of the 600,000 to 700,000 managed Canadian colonies were affected. The impressive Canadian response provides a model for the U.S. transition to post-Africanized beekeeping. While the details will differ, the fundamental lesson from the Canadian experience is that major changes can be accomplished industry-wide, but only if consensus and determination are present. Beekeeping in the United States is at a crossroad, one where opportunities will arise for those flexible enough to explore the path of new management options. The Africanized bee does not have to be a disaster for American agriculture. It can provide an exciting opportunity to revitalize an industry confronted with a serious but not insurmountable challenge.

Lessons to Be Learned

What have we learned from the Africanized honey bee? The most obvious lesson is that foreign insects should be introduced into a new locale only with extreme care. The honey bee is particularly interesting in this regard, because its introduction into the Americas is generally considered to have been highly beneficial for man. It is ironic that in its Africanized form the honey bee has been considered a serious pest. Although tens of millions of dollars have been spent to study, control, and manage this basically beneficial insect, we are nowhere near the end of the havoc wreaked by these bees. Yet beekeepers and bee scientists continue to propose the introduction of various bee races to new habitats, with the unrealistic expectation that foreign bees will somehow have better characteristics than those that are naturally present or have already been introduced.

The Africanized bees have also revealed numerous weaknesses in our system of agricultural regulation and research. The inability of federal and state agencies to cooperate and reach a consensus concerning their individual responsibilities is a serious concern. Further, conflicting interests within commodity groups can and do aggravate the inherent tendency of elected governments to pass the buck, and can contribute to an almost paralyzing inability to respond to pest situations. When personality clashes, funding battles, and scientific disagreements among the "experts" are added to the mix, only chaos can result. Beekeeping in America has, to some extent, been a victim of all these factors. Our responses to Africanized bees have been, and will continue to be, less effective than they would be if cooperation were the order of the day. There is still time to pull together a rational Africanized bee policy, but only if the government, beekeepers, and scientists work together. The action program presented here could form the backbone of such a program.

We also have learned from the Africanized bee how effective the media can be in distorting the facts. The barrage of killer bee stories cascading from the press has served to warp the public's understanding of the problems caused by these bees, and by and large has failed to reveal the fascinating biology involved. Certainly the media have a serious responsibility to report the more dangerous aspects of Africanized bees; but misleading headlines, exaggerated rhetoric, and reporting limited to stinging incidents have not served the public interest. Although there have been some well-done media pieces on these bees, the general tenor of Africanized bee stories has ranged from cute to lurid. Surely science reporting can do better, by paying more attention to the full complement of traits that have made possible the remarkable success of this insect.

Indeed, I would like to believe that our understanding of the marvelous adaptations of the Africanized bees will be their most profound legacy. This is an elegant insect, superbly adapted to its tropical lifestyle, and research into its biology has contributed to a renaissance of research on social insects. New positions at universities, colleges, and government research and regulatory agencies have developed as a direct response to the need for learning more about this bee. Studies of its attributes have made significant contributions to ecology, genetics, behavior, molecular biology, and other fields. Honey bee science has matured from a narrow discipline focused on applied management questions to one that is at the forefront of contemporary biology. Some of today's best young scientists have chosen to work with this complex insect. When all the action programs, controversies, and dilemmas caused by Africanized bees have receded into the past, we will be left with a natural history paradigm that can only contribute to our appreciation of the biological world.

Selected Sources and Readings

Index

Selected Sources and Readings

General

American Farm Bureau Research Foundation. 1986. Proceedings of the Africanized Honey Bee Symposium, Atlanta. Park Ridge, Ill.: American Farm Bureau Research Foundation.

Needham, G. R., R. E. Page, Jr., and M. Delfinado-Baker. 1988. *Africanized Honey Bees and Bee Mites.* West Sussex, England: Ellis Horwood.

Seeley, T. D. 1985. *Honeybee Ecology.* Princeton: Princeton University Press.

Spivak, M., D. J. C. Fletcher, and M. D. Breed. 1991. *The "African" Honey Bee.* Boulder: Westview Press.

Winston, M. L. 1987. *The Biology of the Honey Bee.* Cambridge, Mass.: Harvard University Press.

———. 1992. The Africanized honey bee. *Annual Review of Entomology* 37:173–193.

Winston, M. L., O. R. Taylor, and G. W. Otis. 1983. Some differences between temperate European and tropical African and South American honeybees. *Bee World* 64:12–21.

1. The Creation of a Pop Insect

Nunamaker, R. A. 1979. Being stung by the press. *American Bee Journal* 119:587–592, 646–647, 657.

2. Arrival of the Bees

Goncalves, L. S. 1974. The introduction of the African bees (*Apis mellifera adansonii*) into Brazil and some comments on their spread in South America. *American Bee Journal* 114:414–415, 419.

Kerr, W. 1957. Introducão de abelhas africanas no Brasil. *Brasil Apicola* 3:211–213.

———. 1967. The history of the introduction of Africanized bees to Brazil. *South African Bee Journal* 38:3–5.

Nogueira-Neto, P. 1964. The spread of a fierce African bee in Brazil. *Bee World* 45:119–121.

Portugal, A. V. de. 1971. The central African bee in South America. *Bee World* 52:116–121.

Roubik, D. W. 1989. *Ecology and Natural History of Tropical Bees.* New York: Cambridge University Press.

Schnetler, E. A. 1946. Honey production record. *South African Bee Journal* 21:10.

Smith, M. V. 1960. *Beekeeping in the Tropics.* London: Longmans.

Spivak, M., D. J. C. Fletcher, and M. D. Breed. 1991. *The "African" Honey Bee,* pp. 1–8. Boulder: Westview Press.

3. Temperate and Tropical Honey Bees

Fletcher, D. J. C. 1978. The African bee, *Apis mellifera adansonii,* in Africa. *Annual Review of Entomology* 23:151–171.

Jay, S. C. 1963. The development of honey bees in their cells. *Journal of Apicultural Research* 2:117–134.

Morales, G. 1986. Effects of cavity size on demography of unmanaged colonies of honey bees (*Apis mellifera* L.). M.Sc. thesis, University of Guelph, Ontario.

Otis, G. W., M. L. Winston, and O. R. Taylor. 1981. Engorgement and dispersal of Africanized honeybee swarms. *Journal of Apicultural Research* 20:3–12.

Ratnieks, F., M. Piery, and I. Quadriello. 1991. The natural nest of the Africanized honey bee near Tapachula, Chiapas, Mexico. *Canadian Entomologist* 123:353–359.

Schneider, S., and R. Blyther. 1988. The habitat and nesting biology of the African honeybee *Apis mellifera scutellata* in the Okavanso River delta, Botswana, Africa. *Insectes Sociaux* 35:167–181.

Seeley, T. D. 1978. Life history strategy of the honeybee, *Apis mellifera. Oecologia* 32:109–118.

Seeley, T. D., and R. Morse. 1976. The nest of the honey bee (*Apis mellifera*). *Insectes Sociaux* 23:495–512.

Tribe, G. D., and D. J. C. Fletcher. 1977. Rate of development of the workers of *Apis mellifera adansonii* L. In *African Bees: Their Taxonomy, Biology, and Economic Use,* ed. D. J. C. Fletcher, pp. 115–119. Pretoria: Apimondia.

Winston, M. L. 1979. Intra-colony demography and reproductive rate of the Africanized honey bee in South America. *Behavioral Ecology and Sociobiology* 4:279–292.

Winston, M. L., and S. J. Katz. 1981. Longevity of cross-fostered honey bee workers (*Apis mellifera*) of European and Africanized races. *Canadian Journal of Zoology* 63:777–780.

———. 1982. Foraging differences between cross-fostered honeybee workers (*Apis mellifera*) of European and Africanized races. *Behavioral Ecology and Sociobiology* 10:125–129.

Winston, M. L., J. Dropkin, and O. R. Taylor. 1981. Demography and life history characteristics of two honey bee races (*Apis mellifera*). *Oecologia* 48:407–413.

4. Seasonal Patterns, Swarming, and Absconding

Cosenza, G. W. 1972. Estudo dos enxames de migracão de abelhas africans. *Proceedings of the Brazilian Congress on Apiculture,* pp. 128–129.

Nightingale, J. 1976. Traditional beekeeping among Kenya tribes and methods proposed for improvement and modernisation. In *Apiculture in Tropical Climates,* ed. E. Crane, pp. 15–22. London: International Bee Research Association.

Otis, G. W. 1991. Population biology of the Africanized honey bee. In M. Spivak, D. J. C. Fletcher, and M. D. Breed, *The "African" Honey Bee,* pp. 198–218. Boulder: Westview Press.

Otis, G. W., M. L. Winston, and O. R. Taylor. 1981. Engorgement and dispersal of Africanized honey bee swarms. *Journal of Apicultural Research* 20:3–12.

Schneider, S. S. 1990. Nest characteristics and recruitment behavior of absconding colonies of the African honey bee, *Apis mellifera scutellata,* in Africa. *Journal of Insect Behavior* 3:225–240.

Seeley, T. D., and P. K. Visscher. 1985. Survival of honeybees in cold climates: the critical timing of colony growth and reproduction. *Ecological Entomology* 10:81–88.

Simpson, J. 1958. The factors which cause colonies of *Apis mellifera* to swarm. *Insectes Sociaux* 5:77–95.

Winston, M. L. 1979. Intra-colony demography and reproductive rate of the Africanized honey bee in South America. *Behavioral Ecology and Sociobiology* 4:279–292.

———. 1980. Swarming, afterswarming, and reproductive rate of unmanaged honeybee colonies (*Apis mellifera*). *Insectes Sociaux* 27:391–398.

Winston, M. L., and G. W. Otis. 1978. Ages of bees in swarms and afterswarms of the Africanized honeybee. *Journal of Apicultural Research* 17:123–129.

Winston, M. L., and O. R. Taylor. 1980. Factors preceding queen rearing in the Africanized honeybee (*Apis mellifera*) in South America. *Insectes Sociaux* 27:289–304.

Winston, M. L., J. Dropkin, and O. R. Taylor. 1981. Demography and life history characteristics of two honey bee races (*Apis mellifera*). *Oecologia* 48:407–413.

Winston, M. L., G. W. Otis, and O. R. Taylor. 1979. Absconding behavior of the Africanized honeybee in South America. *Journal of Apicultural Research* 18:85–94.

5. Activities outside the Nest

Breed, M. 1991. Defensive behavior. In M. Spivak, D. J. C. Fletcher, and M. D. Breed, *The "African" Honey Bee*, pp. 279–286. Boulder: Westview Press.

Collins, A. M., and T. E. Rinderer. 1991. Genetics of defensive behavior I. In M. Spivak, D. J. C. Fletcher, and M. D. Breed, *The "African" Honey Bee*, pp. 287–305. Boulder: Westview Press.

Collins, A. M., T. E. Rinderer, J. R. Harbo, and A. B. Bolten. 1982. Colony defense by Africanized and European honey bees. *Science* 218:72–74.

Danka, R. G., R. L. Hellmich II, T. E. Rinderer, and A. M. Collins. 1987. Diet-selection ecology of tropically and temperately adapted honey bees. *Animal Behavior* 35:1858–63.

Danka, R. G., T. E. Rinderer, R. L. Hellmich II, and A. M. Collins. 1986. Foraging population sizes of Africanized and European honey bee (*Apis mellifera* L.) colonies. *Apidologie* 17:193–202.

Frisch, K. von. 1967. *The Dance Language and Orientation of Bees*. Cambridge, Mass.: Harvard University Press.

Hellmich, R. L. II, and T. E. Rinderer. 1991. Beekeeping in Venezuela. In M. Spivak, D. J. C. Fletcher, and M. D. Breed, *The "African" Honey Bee*, pp. 396–408. Boulder: Westview Press.

Nuñez, J. A. 1979. Time spent on various components of foraging activity: comparison between European and Africanized honeybees in Brazil. *Journal of Apicultural Research* 18:110–115.

Pesante, D., T. E. Rinderer, and A. M. Collins. 1987a. Differential nectar foraging by Africanized and European honeybees in the neotropics. *Journal of Apicultural Research* 26:210–216.

———. 1987b. Differential pollen collection by Africanized and European honeybees in Venezuela. *Journal of Apicultural Research* 26:24–29.

Rinderer, T. E., and A. M. Collins. 1991. Foraging behavior and honey production. In M. Spivak, D. J. C. Fletcher, and M. D. Breed, *The "African" Honey Bee,* pp. 219–241. Boulder: Westview Press.

Rinderer, T. E., A. B. Bolten, A. M. Collins, and J. R. Harbo. 1984. Nectar foraging characteristics of Africanized and European honeybees in the neotropics. *Journal of Apicultural Research* 23:70–79.

Rinderer, T. E., A. M. Collins, and K. W. Tucker. 1985. Honey production and underlying nectar harvesting activities of Africanized and European honeybees. *Journal of Apicultural Research* 23:161–167.

Roubik, D. W. 1978. Competitive interactions between neotropical pollinators and Africanized honey bees. *Science* 201:1030–32.

———. 1989. *Ecology and Natural History of Tropical Bees.* New York: Cambridge University Press.

Stort, A. C., and L. S. Goncalves. 1991. Genetics of defensive behavior II. In M. Spivak, D. J. C. Fletcher, and M. D. Breed, *The "African" Honey Bee,* pp. 306–333. Boulder: Westview Press.

Taylor, O. R. 1986. Health problems associated with African bees. *Annals of Internal Medicine* 104:267–268.

6. The Process of Africanization

Buco, S. M., T. E. Rinderer, H. A. Sylvester, A. M. Collins, V. A. Lancaster, and R. M. Crewe. 1987. Morphometric differences between South American Africanized and South African (*Apis mellifera scutellata*) honey bees. *Apidologie* 18:217–222.

Daly, H. V. 1991. Systematics and identification of Africanized bees. In M. Spivak, D. J. C. Fletcher, and M. D. Breed, *The "African" Honey Bee,* pp. 9–41. Boulder: Westview Press.

Daly, H. V., and S. S. Balling. 1978. Identification of Africanized honeybees in the Western hemisphere by discriminant analysis. *Journal of the Kansas Entomological Society* 51:857–869.

Fierro, M. M., M. J. Muñoz, A. Lopez, X. Sumuano, H. Salcedo, and G. Roblero. 1988. Detection and control of the Africanized bee in coastal Chiapas, Mexico. *American Bee Journal* 128:272–275.

Hall, H. G. 1986. DNA differences found between Africanized and European honeybees. *Proceedings of the National Academy of Science, USA* 83:4874–77.

———. 1990. Parental analysis of introgressive hybridization between African and European honeybees using nuclear DNA RFLP's. *Genetics* 125:611–621.

Hall, H. G., and K. Muralidharan. 1989. Evidence from mitochondrial DNA that African honey bees spread as continuous maternal lineages. *Nature* 339:211–213.

Hellmich, R. L. II, T. E. Rinderer, R. G. Danka, A. M. Collins, D. L. Boykin. 1990. Flight times of Africanized and European honey bee drones (Hymenoptera: Apidae). *Journal of Economic Entomology*. In press.

Lobo, J. A., M. A. del Lama, and M. A. Mestriner. 1989. Population differentiation and racial admixture in the Africanized honeybee (*Apis mellifera* L.). *Evolution* 43:794–802.

Rinderer, T. E., and R. L. Hellmich II. 1991. The processes of Africanization. In M. Spivak, D. J. C. Fletcher, and M. D. Breed, *The "African" Honey Bee*, pp. 89–110. Boulder: Westview Press.

Sheppard, W. S., T. E. Rinderer, J. A. Mazzoli, J. A. Stelzer, and H. Shimanuki. 1991. Gene flow occurs between African- and European-derived honey bee populations in Argentina. *Nature* 339:213–215.

Smith, D. R., O. R. Taylor, and W. W. Brown. 1989. Neotropical Africanized honey bees have African mitochondrial DNA. *Nature* 339:213–215.

Sylvester, H. A., and T. E. Rinderer. 1987. Fast Africanized bee identification system (FABIS) manual. *American Bee Journal* 127:511–516.

7. The Latin American Experience

Anonymous. 1972. National Research Council, National Academy of Sciences. Final report of the Committee on the African Honey Bee. Washington, D.C. 95 pp.

Boreham, M. M., and D. W. Roubik. 1987. Population change and control of Africanized honey bees in the Panama Canal area. *Bulletin of the Entomological Society of America* 33:34–38.

DeJong, D. 1984. Africanized bees now preferred by Brazilian beekeepers. *American Bee Journal* 124:116–118.

Goncalves, L. S. 1974. The introduction of the African bees into Brazil and some comments on their spread in South America. *American Bee Journal* 114:414, 415, 419.

Goncalves, L. S., A. C. Stort, and D. DeJong. 1991. Beekeeping in

Brazil. In M. Spivak, D. J. C. Fletcher, and M. D. Breed, *The "African" Honey Bee,* pp. 334–346. Boulder: Westview Press.

Goncalves, L. S., W. E. Kerr, J. C. Netto, and A. C. Stort. 1974. Some comments on the "Final Report of the Committee on the African Honey Bee." Department of Genetics, Faculty of Medicine, Ribeirao Preto, Brazil.

Hellmich, R. L. II, and T. E. Rinderer. 1991. Beekeeping in Venezuela. In M. Spivak, D. J. C. Fletcher, and M. D. Breed, *The "African" Honey Bee,* pp. 396–408. Boulder: Westview Press.

Michener, C. D. 1975. The Brazilian bee problem. *Annual Review of Entomology* 20:339–416.

Nogueira-Neto, P. 1964. The spread of a fierce African bee in Brazil. *Bee World* 45:119–121.

Rodriguez, R. G. 1979. Presencia de la abeja africanizada en Venezuela: comentarios en relacion a su impacto sobre la apicultura y salud publica. Report No. 5, Ministry of Agriculture, Venezuela.

Roubik, D. W., and M. M. Boreham. 1990. Learning to live with Africanized honeybees. *Interciencia* 15:146–153 (contains information on Panama).

Taylor, O. R. 1984. Challenges Africanized bee article. *American Bee Journal* 124:395–396.

Taylor, O. R., and M. D. Levin. 1978. Observations on Africanized honey bees reported to South and Central American government agencies. *Bulletin of the Entomological Society of America* 24:412–414.

Wiese, H. 1977. Apiculture with Africanized bees in Brazil. *American Bee Journal* 117:166–170.

Woyke, J. 1969. African honey bees in Brazil. *American Bee Journal* 109:342–344.

8. Prognosis for North America

Camazine, S., and R. Morse. 1988. The Africanized honeybee. *American Scientist* 76:465–471.

Dietz, A., R. Krell, and F. A. Eischen. 1985. Preliminary investigation on the distribution of Africanized honey bees in Argentina. *Apidologie* 16:99–108.

Dietz, A., R. Krell, and J. Pettis. 1986. The potential limit of survival for Africanized bees in the United States. *Proceedings of the Africanized Honey Bee Symposium,* American Farm Bureau Research Foundation, Atlanta.

———. 1988. Survival of Africanized and European honey-bee colonies confined in a refrigeration chamber. In G. R. Needham, R. E. Page, Jr., M. Delfinada-Baker, and C. E. Bowman, *Africanized Honey Bees and Bee Mites,* p. 237–242. England: Ellis Harwood.

Kerr, W. E., S. L. del Rio, and M. D. Barrionuevo. 1982. The southern limits of the distribution of the Africanized honey bee in South America. *American Bee Journal* 122:196–198.

McDowell, R. 1984. The Africanized honey bee in the United States. Agricultural Economic Report no. 519. Economic Research Service, U.S. Department of Agriculture.

Rinderer, T. E. 1986. Africanized bees: the Africanization process and potential range in the United States. *Bulletin of the Entomological Society of America* 31:222–227.

Robinson, W. S., R. Nowogrodzki, and R. A. Morse. 1989. The value of honey bees as pollinators of U.S. crops. *American Bee Journal* 129:411–423.

Southwick, E. E., D. W. Roubik, and J. M. Williams. 1990. Comparative energy balance in groups of Africanized and European honey bees: ecological implications. *Comparative Biochemical Physiology* 97A:1–7.

Taylor, O. R. 1977. The past and possible future spread of Africanized honeybees in the Americas. *Bee World* 58:19–30.

———. 1985. African bees: potential impact in the United States. *Bulletin of the Entomological Society of America* 31:15–24.

Taylor, O. R., and M. Spivak. 1984. Climatic limits of tropical African honeybees in the Americas. *Bee World* 65:38–47.

Villa, J. D., N. Koeniger, and T. E. Rinderer. 1991. Overwintering of Africanized, European, and hybrid honey bees in Germany. *Environmental Entomology* 20:39–43.

Woyke, J. 1973. Experiences with *Apis mellifera adansonii* in Brazil and Poland. *Apiacta* 8:115–116.

9. Coping with the Bees

Backus, R. 1990. APHIS ends Mexico program; no plans for Texas quarantine. *Speedy Bee* 19:5, 10.

Cobey, S., and T. Lawrence. 1985. Status of the Africanized bee find in California. *American Bee Journal* 125:607–611, 672–675.

Danka, R. G., and T. E. Rinderer. 1986. Africanized bees and pollination. *American Bee Journal* 126:680–682.

Danka, R. G., T. E. Rinderer, A. M. Collins, and R. L. Hellmich II.

1987. Responses of Africanized honey bees to pollination-management stress. *Journal of Economic Entomology* 80:621–624.

Flottum, K. 1989. The Mexican connection. *Gleanings in Bee Culture* 117:228, 230.

Hellmich, R. L. II, and G. D. Waller. 1990. Preparing for Africanized honey bees: evaluating control in mating apiaries. *American Bee Journal* 130:537–542.

Hellmich, R. L. II, T. E. Rinderer, R. G. Danka, A. M. Collins, and D. L. Boykin. 1988. Influencing matings of European honey bee queens in areas with Africanized honey bees. *Journal of Economic Entomology* 81:796–799.

Morse, R. A., D. M. Burgett, J. T. Ambrose, W. E. Conner, and R. D. Fell. 1973. Early introductions of African bees into Europe and the New World. *Bee World* 54:57–60.

Page, R. E., and E. H. Erickson, Jr. 1985. Identification and certification of Africanized honey bees. *Annals of the Entomological Society of America* 78:149–158.

Report on the African bee barrier program. 1986. *Gleanings in Bee Culture* 114:622, 623, 636, 637, 645.

Rinderer, T. E., J. E. Wright, H. Shimanuki, F. Parker, E. Erickson, and W. T. Wilson. 1987. The proposed honey-bee regulated zone in Mexico. *American Bee Journal* 127:160–164.

Stibick, J. N. L. 1984. Animal and plant health inspection service strategy and the African honey bee. *Bulletin of the Entomological Society of America* 29:22–26.

Texas Africanized Honey Bee Management Plan. 1989. Texas Africanized Honey Bee Commission.

Yeutter, C. 1990. U.S. Department of Agriculture policy on Africanized honey bees. *Speedy Bee* 19:5.

Index